W0055277

FREQUENCY CURVES
AND
CORRELATION

FREQUENCY CURVES
AND
CORRELATION

by

WILLIAM PALIN ELDERTON

K.B.E., F.I.A., F.F.A.

FOURTH EDITION

CAMBRIDGE
UNIVERSITY PRESS

CAMBRIDGE UNIVERSITY PRESS
Cambridge, New York, Melbourne, Madrid, Cape Town,
Singapore, São Paulo, Delhi, Tokyo, Mexico City

Cambridge University Press
The Edinburgh Building, Cambridge CB2 8RU, UK

Published in the United States of America by Cambridge University Press, New York

www.cambridge.org
Information on this title: www.cambridge.org/9781107601291

© Cambridge University Press 1906

This publication is in copyright. Subject to statutory exception
and to the provisions of relevant collective licensing agreements,
no reproduction of any part may take place without the written
permission of Cambridge University Press.

First edition 1906
Second edition 1927
Third edition 1938
Fourth edition 1953
First paperback edition 2011

A catalogue record for this publication is available from the British Library

ISBN 978-1-107-60129-1 Paperback

Cambridge University Press has no responsibility for the persistence or
accuracy of URLs for external or third-party internet websites referred to in
this publication, and does not guarantee that any content on such websites is,
or will remain, accurate or appropriate.

PREFACE TO SECOND EDITION

This book arose out of an attempt, published in 1903, to use Professor Pearson's system of frequency curves for the graduation of Mortality Tables. The subject was then unfamiliar to actuaries, and the Institute of Actuaries encouraged me to write a book on the subject, arranged for its publication and relieved me of any expense in connection with it. My gratitude is not only for the broad-mindedness with which a professional body approached recent research, but also for the help and encouragement given to a young, untried and inexperienced member of the profession. Nor does it end there: for when the original edition was nearly exhausted the Institute generously handed over the copyright and left me with a free hand as regards the future.

In dealing with frequency curves and curve fitting, new matter has been brought in and the order of treatment of the curves has been altered so that main types are less likely to be confused with the transition and minor types. A chapter is devoted to a comparison of the Pearson curves with the series suggested for use by Edgeworth in this country and by many continental writers. The chapters on correlation, contingency, etc. have been largely rewritten and a new chapter on partial correlation added.

The book as it now stands assumes that the reader is familiar with the *Primer of Statistics* or some other very elementary book. It demands no mathematical knowledge beyond that required for the first examination of the Institute of Actuaries or the Intermediate Examination for the B.Sc. of London University. The subject is, however, statistical and arithmetical, and examples must be worked out if the methods and principles are to be mastered. The reader who goes through a book on a practical subject and does not work out examples is as certain to encounter imaginary and miss real difficulties as he is to fail to obtain any satisfactory knowledge of the subject.

Even if a reader does not possess the mathematical equipment indicated, he can use frequency curves and correlation reasonably without it, for the fact that a curve he has found agrees with the statistics from which the moments were obtained is a proof that, in the particular case, he has obtained proper values for the constants, even though he has not followed the mathematical reasoning leading to the equations. It must not be inferred that belief without proof is advisable, but that it is unwise for a practical man to put aside a practical subject which he can test practically merely because he cannot follow some of the proofs. There is another class of statistical students whose wants may be mentioned. I refer to those who have little need to study graduation and curve fitting in detail, but require a knowledge of correlation, probable errors, etc. For the sake of these readers an abridged reading is suggested in the Appendix.

Frequency curves, correlation and sampling form a subject in which there is still a great deal to be done, notwithstanding the progress that has been made in recent years. Much of this work has been highly mathematical, especially when it deals with certain small samples or with attempts to find mathematical expressions for skew correlation surfaces. These aspects lie outside such a book as this, and, even if we neglect them, we may still say that there are few subjects that offer a richer field for original work. In this field the reader will find that during the past thirty years we are indebted to Professor Karl Pearson and his school for much of the work that has proved a success in practice, and anyone writing on the subject for practical men is bound to follow in his footsteps. Only those who become interested in the subject and study Professor Pearson's original work will appreciate the great extent of his contribution to statistical science.

I hope that the numerical examples in this book, and similar arithmetical work done elsewhere, may tend to show that actuarial statistics can be examined in the same way as the statistics of biology, anthropology or sociology. May not such work add some links to the chain of continuity and indicate a wider law than an actuary studying his own subject exclusively might be led to suspect?

As will be readily appreciated, I am chiefly indebted to Professor Pearson, but the indebtedness is of a kind for which it is impossible to offer formal thanks; such thanks would, at their best, fail to express the sense of gratitude which prompted them.

The revision of the book has been a reminder of much kind help received in connection with the first edition from Mr G. J. Lidstone and Mr John Spencer, both of whom read the work in a somewhat different form in MS. and made many valuable suggestions, and from Messrs S. Adlard and R. L. Elderton, who then spent much time in reading proofs and suggested difficulties that would probably arise and ways of removing them. In connection with the new edition, Mr H. B. Smither has helped with some of the calculations, and both he and Mr H. T. Adlard have read the book in proof; help for which I am, indeed, grateful. At many stages in the work my sister, Miss Ethel M. Elderton, has come to my aid; and, bearing in mind her experience in teaching the subject as well as her practical work, it would have been better if the book had been hers and not mine: any improvement in this edition is probably hers already.

W. P. E.

19 COLEMAN STREET
LONDON, E.C. 2

July 1927

PREFACE TO THIRD EDITION

The book has been altered in many respects, and Chapters x, xi and xii and some of the Appendices have been rewritten.

The notation for moments in the earlier editions has been retained. Some writers find it helpful to use distinct symbols for the "theoretical moment" and the "adjusted statistical moment". In practical curve fitting the two are equated. The notation I use treats the latter as identical with the former. Readers of other work, and especially of continental work, must bear in mind these and other differences in notation.

I am most grateful to Professor E. S. Pearson for the help and advice he has given me so generously and sympathetically. It is also a pleasure to thank Mr H. Latham Seal for many suggestions and him and Mr H. J. Tappenden for much help in connection with the proofs.

I hope these kind friends will not be thought to be in any way responsible for my shortcomings.

W.P.E.

October 1937

PREFACE TO FOURTH EDITION

The third edition went out of print a few years ago. There is still some demand for the book and, as reprinting presented difficulties, it has been reproduced photographically. Substantial revision was therefore out of the question. I might otherwise have given examples of graduations by the lognormal curve and said more about the method of translation especially as Professor J. F. Steffensen has pointed out to me that if it be used for the exposed to risk with a Pearson Type III curve putting $f(x) = c^x$ the approximate law of survivorship given by me in 1932 would hold good if the mortality follows Makeham's "law". I should also have been glad to add a further warning about the assumptions of linear regression and skew correlation surfaces more or less on the lines of the paper I wrote for the Royal Statistical Society, Vol. cviii. I have been able to correct a few arithmetical slips but I must confess to one that I have left unaltered. On p. 68 (last line) μ_4 should be 65·042264; the effect on the graduation is small and many trifling alterations in subsequent pages would have been necessary.

W. P. E.

CONTENTS

(xi)

CHAPTER I

INTRODUCTORY

1. The ordinary treatment of probability begins with the assumption that the chance that a certain event will occur is known, and proceeds to solve the problems that arise from the combination of events or the repetition of a particular experiment; it proves that a certain result is more likely to occur from experiment than any other, that a result based on a limited number of trials is unlikely to differ greatly from the expected result, and that the proportional deviation from the most probable result will generally decrease as the number of trials is increased.

Experiments can easily be made to show that the theoretical method leads to results which can be realised in practice when the probabilities can be estimated accurately beforehand; for example, various trials have been made with coin tossing in which it has been found that if five coins are tossed together and the number of them coming down "heads" is recorded, then the distribution of the cases will agree with the binomial expansion $(\frac{1}{2} + \frac{1}{2})^5$ as the ordinary theory leads us to expect. Sequences of "heads" or "tails" form a series approximating to the geometrical progression with a common ratio of $\frac{1}{2}$, and the drawing of cards from a pack gives a result closely agreeing with the numbers that theoretical work suggests.

2. It frequently happens, however, that the probabilities are not known, and it is impossible to tell whether we are dealing with an experiment like coin tossing or sequences or card-drawing; in fact, the only thing known is the distribution of the number of cases into certain groups, and in these circumstances the inverse problem of tracing the theoretical series to which the statistics approximate may become an important matter. The difficulty of the subject is increased because statistics do not give the theoretical distribution exactly, and it is impossible to tell where the differences between the actual and theoretical results lie. To make the position clearer it will be well to restate the problem and ask whether it is possible to find the theoretical series to which a series, resulting from a statistical experiment, approximates. It may be difficult, perhaps impossible, to trace the probabilities corresponding to a given case, but yet practicable to form a reasonable opinion of the series of numbers that might be reached if the experiment could be repeated an infinite number of times. On turning to the reasons which make it advisable to find this ideal result to which statistics approach, it will be seen that the exact elementary probabilities are not of supreme importance, and a reasonable representation of the series is of far greater practical value. We notice that one of the first objects of a statistician or an actuary dealing with statistical work is to express the observations in a simple form so that practical conclusions can be easily drawn from the figures that have been collected. If the available statistics fall naturally into fifty or sixty groups, he has to decide how they can be arranged to bring out the important features of the problem on which he is working; whereas if he can find a few numbers closely connected with the original series which can be used as an index to the whole, he can then give the result in a way that might assist comparison with similar statistics, and enable others who have to deal with the facts to appreciate the whole distribution more readily than they could do if it remained in its original form. The statistician has also to supply approximate values for intermediate terms when only a few can be obtained from his experience, or complete or continue a series when only a part of it is known. In many cases he has to keep the same terms as in his original series, but remove the roughnesses of material due to limitations in the number of cases available for his investigation; that is, he has to graduate his data.

3. In reality these objects are much alike, for if the statistical tables can be represented by an algebraic or transcendental formula, we can replace the whole series of numbers

by a few values (the constants in the formula) which, if we deal systematically with the distributions we meet, facilitate comparison or enable us to supply missing terms, while the roughness of the original material can be removed by making a suitable formula represent the original statistics as nearly as possible. If a formula is based on theoretical considerations, it may also give a solution of the problem in probabilities mentioned at the outset, and we see that both the practical and theoretical requirements can be dealt with at the same time, for the smooth series sought by the theoretical student is the same thing as the formula required for practical work.

4. The advantages of any system of curves depend on the simplicity of the formulae and the number of classes of observations that can be dealt with satisfactorily, for a complicated expression is very little improvement on the original groups of statistics, and a system which is not capable of general application leaves the statistician in difficulties whenever it breaks down. One other thing is necessary; if a formula is known to be a suitable one, there must be some method of finding the arithmetical constants that will give a good agreement in the particular case. Such a method, if it is to be of practical use, must be simple, reliable and capable of general and systematic application.

A broad idea of the objects to be accomplished ought to be kept clearly before the mind; they are likely to be forgotten because of the large amount of detail necessarily connected with the subject. It is also important because the advantages of systematic treatment are often overlooked, and short cuts and rough and ready methods are adopted to the detriment of the work, and formulae having no scientific basis and having no connection with others suitable to similar cases are sometimes used in rather haphazard fashion by statisticians. The consequence is that generalisation is impossible, and where a law might be found one can see little but a great variety of attempts by energetic workers to reach their own conclusions regardless of the value of comparative statistics.

CHAPTER II

FREQUENCY DISTRIBUTIONS

1. If statistics are arranged so as to show the number of times, or frequency with which, an event happens in a particular way, then the arrangement is a frequency distribution. Although some of our results will be of wider applicability, we shall generally confine our attention to these distributions.

2. It is necessary to have a name for the formula used to describe such distributions, and the term "frequency-curve" has been adopted for the purpose. The geometrical progression which describes the number of sequences in any direct experiment, such as coin tossing or dice throwing, is, in the limit, a frequency-curve, the equation to which is $y = Na^x$.

3. Some distributions give the number of cases falling in a certain group of values of the independent variable, while others (e.g. Example V of Table I) give the number of cases for an exact value. In the former case the exact values of the independent variable to which the groups correspond must be considered; for instance, "exposed to risk at age x" includes those from $x - \frac{1}{2}$ to $x + \frac{1}{2}$, but the number of deaths at duration n those from n to $n + 1$. When statistics are represented graphically, effect should be given to these differences, and, to bring out the points a little more clearly, the diagrams on pp. 5 and 6 have been prepared. The drawings of distributions, such as those in the diagrams, are called frequency polygons or histograms.

4. When statistics give the number of cases for an exact value of the independent variable, it is simple to plot them in a diagram by drawing ordinates and joining their tops, but in the case of groups of values there is a little complication, for we can either draw a rectangle standing on the entire base

(4)

(Example II of diagram) or put in ordinates at the middle points of the bases and then join their tops (Example III). The former method gives the correct idea of the amount of information conveyed by the statistics, but, for some purposes (e.g. for seeing the possible shape of the curve), the latter is more convenient, though it is open to technical objection. Cases such as Examples I and IV are best expressed by the kind of drawing given, while Example III though open to

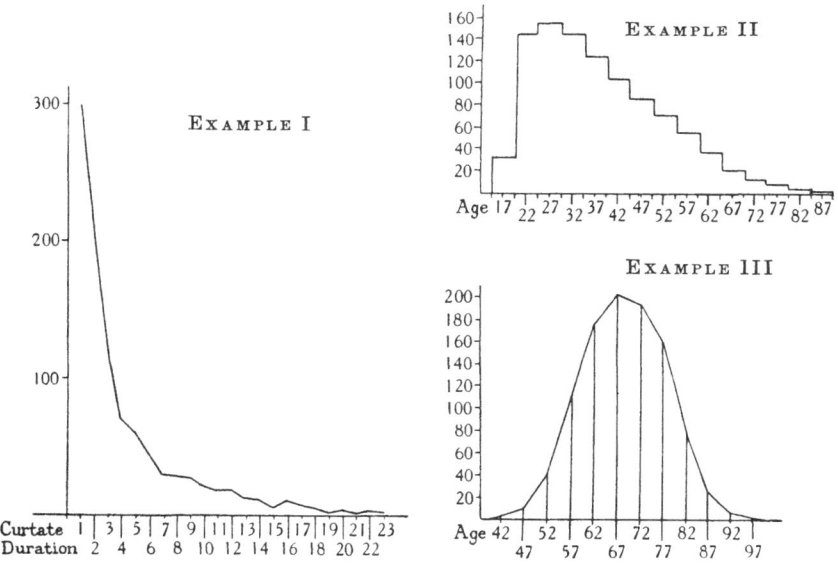

technical objection gives a better indication to most people of the shape of the actual distribution than a block diagram.

5. The reader is no doubt already familiar with the fact that statistics tend towards a smooth series as the total number of cases is increased, and from this it can be seen how naturally practical statistics lead to the conception of a frequency-curve to describe the smooth distribution that would be obtained if an infinite supply of homogeneous material were available for investigation. In other words, such curves would give an

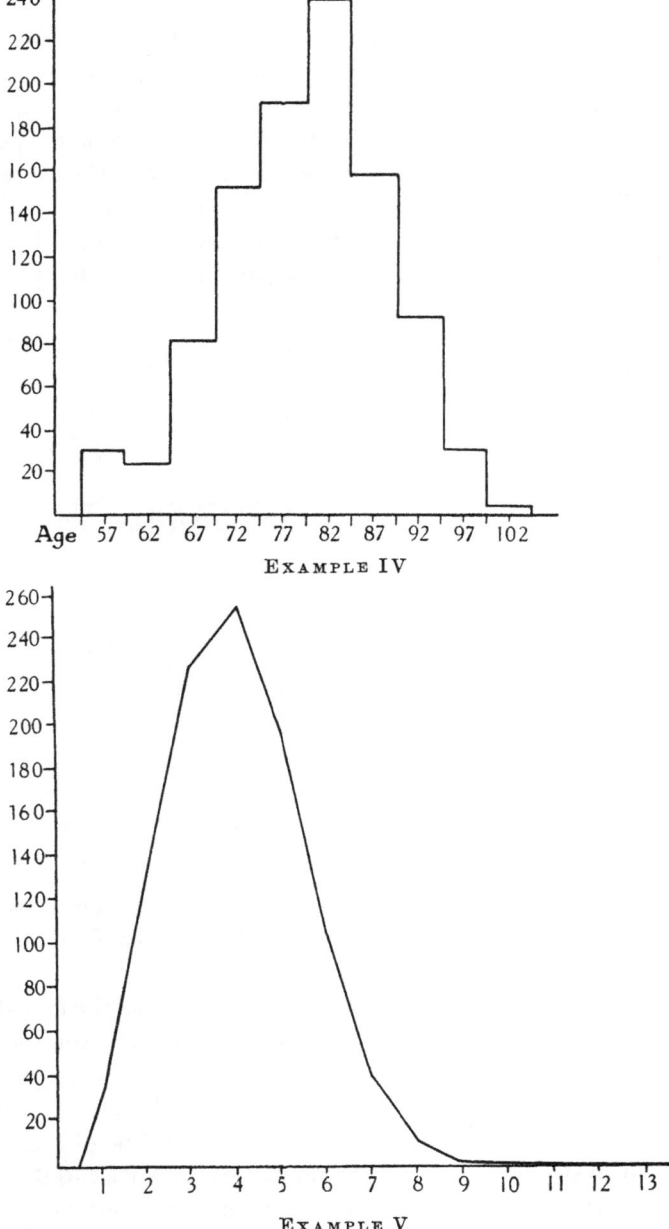

Age   57   62   67   72   77   82   87   92   97   102

Example IV

Example V

(6)

approximation to the total "population" of which the particular case investigated was a sample.

TABLE I

EXAMPLE I		EXAMPLE II		EXAMPLE III	EXAMPLE IV	EXAMPLE V	
Curtate durations	Withdrawals with monthly incidence "0" in year of exit, *Principles and Methods* (p. 92)	Ages	Exposed to risk of sickness (Watson, *M.U. Tables*, p. 19)	Existing at close of observations Without Profit "Old" Assurances	Existing at close of observations "Old" Annuities (females)	Terms of the expansion of $1000(\frac{3}{4}+\frac{1}{4})^{12}$	No. of term
1	308	–19	34	...	...	32	1
2	200	20–24	145	...	...	127	2
3	118	25–29	156	...	...	232	3
4	69	30–34	145	...	...	258	4
5	59	35–39	123	...	...	194	5
6	44	40–44	103	3	...	103	6
7	29	45–49	86	9	...	40	7
8	28	50–54	71	42	...	11	8
9	26	55–59	55	111	29	2	9
10	21	60–64	37	176	23	1	10
11	18	65–69	21	200	81	...	11
12	18	70–74	13	193	151	...	...
13	12	75–79	7	160	192	...	...
14	11	80–84	3	73	239	...	...
15	5	85–89	1	26	157	...	...
16	11	90–94	...	6	93	...	...
17	7	95–99	...	1	29	...	...
18	6	100–	...	...	6	...	...
19	1	...	...	...	...	...	...
20	3	...	...	...	...	...	...
21	1	...	...	...	...	...	...
22	3	...	...	...	...	...	...
23	2	...	...	...	...	...	...
...	1,000	...	1,000	1,000	1,000	1,000	...
True total	1,308	...	2,995,724	2,674	172	...	...
Mean	4·182	...	37·8750	68·485	79·400	3·998	...
Standard deviation	4·1996	...	2·76810	1·771288	1·774894	1·46215	...
Type	I	...	I	II	VII	...	...

6. It may be noticed that a frequency-curve can be interpreted to give a frequency corresponding to every value of the independent variable along the whole range of the distribution, and will not restrict us to a few more or less arbitrary groups as is necessary with actual statistics. The binomial series and geometrical progression do the same when we imagine we are dealing with something that can be divided into a very large

number of groups. Thus, if we mix a large quantity of sand of two colours and take out a fixed quantity of the mixture and record the number of grains of sand of either colour in each drawing, we should obtain a continuous curve from a large number of trials.

7. We will now define some important functions. When a distribution is arranged according to the progressive values of a variable characteristic, e.g. duration, age, etc., the average value of that characteristic (not the average of the frequencies) is called the *mean* of the distribution, and is given by

$$\frac{f_a \times a + f_b \times b + f_c \times c + \ldots + f_n \times n}{f_a + f_b + f_c + \ldots + f_n}$$

where f_r is the frequency corresponding to the value r of the variable; thus, in Example I, 200 is the frequency corresponding to 2. If we assume infinitesimal increments, the mean is given by

$$\int f_x \times x \, dx \Big/ \int f_x \, dx$$

where the limits of the integral will be such as to cover the whole distribution. The mean could also be described as the position of the ordinate through the centre of gravity of the distribution (centroid vertical); this may be of help to some readers.

8. The *mode* is the characteristic that occurs most frequently; in other words, it is the position of the maximum ordinate. We cannot tell from the rough statistics which ordinate is greatest and the mode can therefore only be determined approximately until the law connecting the various groups, i.e. the frequency-curve, is known

9. Now since an equation or curve might be used for several distributions, one given according to age, a second of a different subject according to duration, a third according to sums assured, and so on, we must have a standard of reference based

(8)

on the distribution itself. For this purpose a function known as the *standard deviation* is used. It is given by

$$\sqrt{\left\{\frac{f_a a'^2 + f_b b'^2 + \ldots + f_n n'^2}{f_a + f_b + \ldots + f_n}\right\}}$$

where a', b', ... n' are the distances from the mean. In the form of integrals the standard deviation is

$$\sqrt{\left\{\int f_x \times x^2\, dx \Big/ \int f_x dx\right\}}$$

where x is measured from the mean.

The *standard deviation* measures the way the frequencies are distributed in terms of the unit of measurement. As the frequencies farthest from the mean are multiplied by the largest values of x, a large standard deviation shows that the frequency distribution spreads out from the mean, while a small standard deviation shows that the frequency is closely concentrated about the mean. In considering the relative sizes of standard deviations, it is necessary to bear in mind the unit of measurement, because, if a given distribution is arranged in two series, first, according to years of age, and then in quinquennial age groups, the standard deviation will be five times as large in the latter case as it is in the former. This can be seen at once by comparing the two expressions

$$\sqrt{\left\{\int f_x x^2\, dx \Big/ \int f_x dx\right\}} \quad \text{and} \quad \sqrt{\left\{\int f_x (5x)^2\, dx \Big/ \int f_x dx\right\}}$$

The latter is obviously five times the former. The values of the standard deviations are given in Table I for each case. The diagram on p. 11 shows two curves having the same mean B and approximately the same area, but the dotted curve has the larger standard deviation because it spreads out more on each side of the mean.

The reader will notice from the algebraic expressions given above that the mean, mode and standard deviation are not dependent on the number of cases (i.e. on the absolute size of the curve), but merely on the way they are distributed (i.e. on the proportionate numbers or the shape of the curve). The standard deviation measures the "spread" or "scatter" of the statistics from the mean.

10. An examination of frequency distributions (see Table I and pp. 5 and 6) shows that most of them start at zero, gradually rise to a maximum, and then fall sometimes at a very different rate. If the rise and fall are at the same rate, distribution will be symmetrical about the mean, which must then coincide with the mode. The difference between the mean and mode is therefore a function of the *skewness* or deviation from symmetry. In order to get a satisfactory measure, the spread of the material must be taken into account, and this leads us to measure skewness by the distance between mean and mode divided by standard deviation. If the mean is on the left-hand side of the mode when the statistics are plotted out in diagram, this function will be negative, and to remember the sign it is convenient to write:

$$\text{Skewness} = \frac{\text{Mean} - \text{Mode}}{\text{S.D.}}$$

The diagram on p. 11 will help to show the rationale of the measure for skewness. It gives two curves having the same mean B and the same mode A, but with different standard deviations, and it is clear that the dotted curve, with its larger standard deviation, is more nearly symmetrical than the other curve.

11. We may summarise these functions by saying that the mean and mode fix the position of the curve on the axis; the standard deviation shows how the material is distributed about the mean, and the skewness shows the amount of the deviation from symmetry exhibited by the material.

These preliminary definitions will be sufficient for our present purpose, but the functions defined will be more easily understood when their actual connection with the practical work of curve-fitting has been studied. A student working at the subject for the first time should plot out several distributions on cross-ruled paper, in order to familiarise himself with their nature and appearance. He should calculate and insert the means in the diagrams, but should not attempt to calculate standard deviations until he knows something of the method of moments.

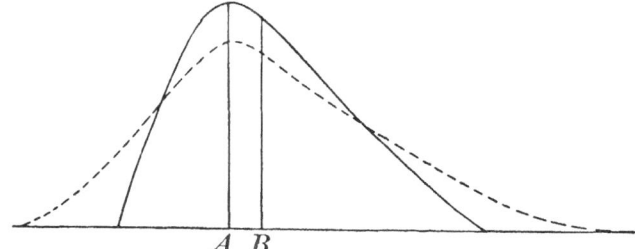

A B

12. Up to this point we have defined our statistics as frequencies, that is, as a number of cases grouped together as alike either because they are actually alike in the sense of Example V or because the statistics throw them up in comparatively narrow groupings as in Examples II, III and IV. When, however, we are tabulating our experience we have to deal with individual observations and they are grouped subsequently. From this point of view if there are N observations we may call them o_1, o_2, o_3, ... o_N, where o_1 may stand for the first observation and may be (see Example III) one of the 200 existing in the 65–69 group. It might be a case "existing" at age 66·12, and o_2 might be "existing" at age 73·72, o_3 at 42·26, o_4 at 67·37 and so on. Then the mean is

$$\frac{1}{N}\sum_{i=1}^{N}o_i$$

CHAPTER III

METHOD OF MOMENTS

1. Before we proceed to deal with suitable forms for use as frequency-curves, it will be well to see if some method of applying them to statistical examples can be found, for it is clearly useless to suggest a curve and have no way of using it. We require, therefore, a general method by which a given formula can be fitted to a particular statistical experience, and may be applied to any expression (for instance, Makeham's formula for the force of mortality) on which we may have decided as the basis of graduation. The first point to be noticed in searching for a method is that if there are n constants in the formula, we must form n equations between the formula and the statistics. Thus, if we have three terms, say, $y = 20$, 40, and 88, when $x = 1$, 2, and 3 respectively, and wish to use the curve $y = a + bx + cx^2$ to describe them, we can, of course, find values of a, b and c so that each item is exactly reproduced by equating as follows:

$$a + b + c = 20$$
$$a + 2b + 2^2c = 40$$
$$a + 3b + 3^2c = 88$$

But if we have a fourth term $y = 96$ when $x = 4$, and use the values of a, b, and c found from the three equations just given, we should find that when $x = 4$, $y = 164$. This suggests that when there are more terms in the statistics than there are constants, the equations must be formed by using all the terms, not by selecting from them. The graduating curve will not necessarily reproduce exactly any of the observations, but will run evenly through the roughnesses of the observed facts so as to represent their general trend.

2. Let $a_1, a_2, a_3, \ldots a_n$ be n terms to be graduated; then, if the series were perfectly smooth and followed a known law, each term could be reproduced exactly by, say, b_1, b_2, b_3, b_n, where $a_1 = b_1, a_2 = b_2, a_3 = b_3, \ldots$ and $a_n = b_n$. Now, if we consider the two series (the a's and the b's), we see that since each term is reproduced exactly

$$\sum_{r=1}^{n} a_r = \sum_{r=1}^{n} b_r \quad \text{and} \quad \sum_{r=1}^{n} c_r a_r = \sum_{r=1}^{n} c_r b_r$$

where c_r is a numerical coefficient.

This suggests a possible method to apply when each term cannot be reproduced exactly. The total of the graduated figures must be made equal to the total of the ungraduated, and the further equations necessary for finding the unknown constants must be formed by multiplying the various terms by different factors and similarly equating the sums of the graduated and ungraduated products, i.e. $\Sigma c_r a_r = \Sigma c_r b_r$. It still remains to decide the best form to be given to c_r, and the mean being equal to

$$\frac{a_1 + 2a_2 + \ldots + na_n}{a_1 + a_2 + \ldots + a_n}$$

suggests that $c_r = r$ should give one reasonable equation. Again, since we shall have to use some function of r which, when applied to the graduation formula, will give an integrable form (otherwise we cannot make an equation between $\Sigma c_r b_r$ and $\Sigma c_r a_r$), the powers of r suggest themselves as convenient when integration by parts is attempted. If, therefore, we write $c_r = r^t$ and give t successively the values $0, 1, 2, \ldots$, we can obtain as many equations as we require, and from the first two of them we find, successively, the area and the mean, which will be the same in the graduated and ungraduated figures.

This method is known as the Method of Moments (cf. moments of inertia), and experience has shown that it is a satisfactory method of fitting a curve to an actual statistical experience. Confirmation on the theoretical side has been produced, and while it is possible to invent other methods of fitting particular curves, as has been done by actuaries in connection with Makeham's law of mortality, no better general method has been produced (see Appendix V, for note on other methods).

3. Applying the method to solve the three equations given above, we have

$$(a+b+c)+(a+2b+2^2c)+(a+3b+3^2c) = 20+40+88$$
$$(a+b+c)+2(a+2b+2^2c)+3(a+3b+3^2c) = 20+2\times40+3\times88$$
$$(a+b+c)+2^2(a+2b+2^2c)+3^2(a+3b+3^2c) = 20+2^2\times40+3^2\times88$$

or
$$3a+\ \ 6b+14c = 148$$
$$6a+14b+36c = 364$$
$$14a+36b+98c = 972$$

These equations will give the same result as those from which they were formed, because each of the three terms can be graduated exactly; but if we introduce the fourth term, $x = 4$, $y = 96$, we can modify the moment method by adding a fourth term to each equation given above and obtain

$$4a+\ \ 10b+\ \ 30c = \ \ 244$$
$$10a+\ \ 30b+100c = \ \ 748$$
$$30a+100b+354c = 2508$$

The solution of these equations gives

$$a = -23{\cdot}0$$
$$b = \ \ \ 42{\cdot}6$$
$$c = -\ \ 3{\cdot}0$$

or
$$x = 1 \qquad y = 16{\cdot}6$$
$$x = 2 \qquad y = 50{\cdot}2$$
$$x = 3 \qquad y = 77{\cdot}8$$
$$x = 4 \qquad y = 99{\cdot}4$$

This is a very simple example, but it will probably help to show the way results are reached, and will serve as a foundation for what follows.

4. The nth moment of a particular frequency is defined as the product of the frequency and the nth power of the distance of the frequency from the vertical about which moments are being taken; or the nth moment of any *ordinate y* of a frequency-curve about the vertical through a point distance x from it, is yx^n, and the nth moment of the whole distribution treated as a series of ordinates is $y_1 x_1^n + y_2 x_2^n + ...$, where $y_1 + y_2 + ...$ is the total frequency. Thus, in Example V, the third moment of the frequency 40 for term 7 about the vertical through 3 is $40 \times (+4)^3$.

5. If the ordinates are known, we can calculate the moment for them immediately by multiplying the frequencies by the powers of the distances between them and the vertical about which the moments are required and then adding the results, care being taken to give the distances their proper signs. If areas are given, an approximation is made by assuming them to be concentrated about the ordinates at the middle points of the bases on which they stand. The columns after the third in Table II show the calculation of the moments about the vertical through age 77 for Example IV of Table I, on the assumption that the frequencies are concentrated at the middle points of the bases.

The unit of grouping has been taken as 5 years, and if, as is often convenient, we assume the total frequency to be unity, the totals will have to be divided by 1000. We should generally deal with the actual numbers that occur, but as they have been given in Table I as the distribution of 1000 cases, it will be better to use them in that way in the present case. The numbers -4, -3, ... in col. (3) show the distances from age 77 in terms of the unit of grouping. The centre of any other group would have done almost as well as 77; it is convenient to choose the arbitrary origin so that it is near the mean of the distribution. This makes easier the calculation of the moments about the mean (a result frequently required), and enables the calculator to get a rough check on these moments by comparing them with those about the arbitrary origin. The cols. (4)–(7)

are sufficiently explained by their headings; they are formed successively and checked by multiplying f by s^4, the values of s^4 being taken from a table of the powers of the natural numbers.

TABLE II

Central age of group x	Frequency f	$(x-77)/5$ $=s$	$f \times s$	$f \times s^2$	$f \times s^3$	$f \times s^4$
(1)	(2)	(3)	(4)	(5)	(6)	(7)
57	29	-4	116	464	1,856	7,424
62	23	-3	69	207	621	1,863
67	81	-2	162	324	648	1,296
72	151	-1	151	151	151	151
77	192	0	$\overline{-498}$	...	$\overline{-3,276}$	...
82	239	1	$\overline{239}$	239	$\overline{239}$	239
87	157	2	314	628	1,256	2,512
92	93	3	279	837	2,511	7,533
97	29	4	116	464	1,856	7,424
102	6	5	30	150	750	3,750
Totals	1,000	...	$\dfrac{+978}{+480}$	3,464	$\dfrac{+6,612}{+3,336}$	32,192

NOTATION FOR MOMENTS

$N = $ total frequency.

$\nu_n = n$th unadjusted statistical moment about mean.

$\nu_n' = n$th unadjusted statistical moment about any other point.

$\mu_n = n$th moment from curve about mean.

$\quad = n$th adjusted statistical moment about mean.

$\mu_n' = n$th moment from curve about other point.

$\quad = n$th adjusted statistical moment about other point.

NOTE. ν, ν', μ and μ' always refer to a total frequency of unity.

The arithmetical work may be checked in other ways; for instance, instead of checking the final column by multiplying each term by the appropriate value of x^4, we can form a new column $(x+1)^4 f$, which is the same thing as

$$x^4 f + 4x^3 f + 6x^2 f + 4xf + f$$

The total of this new column can therefore be used to give a check on the multiplication and addition. In the numerical example (Table II) we should have $29 \times (-3)^4$; $23 \times (-2)^4$, etc., 6×6^4; the total of such a column is 69,240, which agrees with the totals of cols. (2)–(7) in the following way:

$$
\begin{array}{rcr}
& & 32{,}192 \\
4 \times 3{,}336 & = & 13{,}344 \\
6 \times 3{,}464 & = & 20{,}784 \\
4 \times \quad 480 & = & 1{,}920 \\
& & 1{,}000 \\
\hline
& & 69{,}240
\end{array}
$$

Helpful tables (Powers and Fourth moments) will be found in *Tables for Statisticians and Biometricians* edited by K. Pearson (Cambridge University Press). I shall in future refer to this book as *Tables for Statisticians*. A student can manage without these volumes, but at some expense of trouble.

6. It has so far been assumed that moments can be calculated about any point, but it is frequently inconvenient to do so; for if we had required them about age 79·4, we should have had to multiply by the powers of $(57 - 79·4)/5$, of $(62 - 79·4)/5$ and so on, and it is quite clear that the labour would have been very great. In such a case we can, however, take the moments about any other more convenient point, and then modify them in the following way:

Let the distance between A, about which the moments are known, and B, about which they are required, be $+d$; thus, if we want moments about 25·7 and have found them about 25, d is ·7; if we had found them about 26, d would have been $-·3$.

Then, if the distance of any ordinate y_r from A is X_r, and from B is x_r, then

$$x_r = X_r - d \quad \text{and} \quad x_r^n = (X_r - d)^n$$

Now, the nth moment of the whole distribution treated as a series of ordinates is $\Sigma y_r X_r^n$ about A, and $\Sigma y_r x_r^n$ about B; so we have

$$
\begin{aligned}
\nu_n'' = \Sigma y_r x_r^n &= \Sigma y_r (X_r - d)^n \\
&= \Sigma[y_r \{X_r^n - n d X_r^{n-1} + \ldots + (-1)^n d^n\}] \\
&= \nu_n' - n d \nu_{n-1}' + \frac{n(n-1)}{2!} d^2 \nu_{n-2}' - \ldots \qquad \ldots\ldots(1)
\end{aligned}
$$

where v_n'' is written for the nth moment about B, and v_n' the nth moment about A.

Instead of (1) we may proceed as follows:

$$v_n' = \Sigma y_r X_r^n = \Sigma[y_r(x_r+d)^n]$$
$$= v_n'' + n\,d\,v_{n-1}'' + \frac{n(n-1)}{2!}d^2 v_{n-2}'' + \dots$$

$\therefore \qquad v_n'' = v_n' - n\,d\,v_{n-1}'' - \frac{n(n-1)}{2!}d^2 v_{n-2}'' - \dots \qquad \dots\dots(2)$

There is little to choose between these two formulae, and of course they give identical results.

7. We will now apply them to work out the moments about the centroid vertical (i.e. vertical through the mean) for the example in Table II. The distance of the mean from any point is

$$\frac{\Sigma(X_r y_r)}{\Sigma y_r} = \frac{\Sigma(X_r y_r)}{N}$$

where N is the total frequency; or we may say that the distance of the mean from any point is the first moment of the distribution about the vertical through that point. It follows that the first moment about the centroid vertical is zero, so that if such moments are required the term involving v_1'' in (2) is zero. When we deal with frequency-curves we shall see that we generally require moments about the centroid vertical and in designating them we shall leave out the dashes and use v.

8. The arithmetical work is as follows:

The totals in cols. (4)–(7) are divided by the number of observations (total of col. (2)), and the quotients are the moments (v') about 77. The moments are dealt with as having reference to a case where unity is the total frequency, i.e. proportional, not actual, frequencies are dealt with.

$$v_1' = \cdot480 \qquad v_2' = 3\cdot464$$
$$v_3' = 3\cdot336 \qquad v_4' = 32\cdot192$$

The value of v_1' gives the mean age $= 77 + 5 \times \cdot480 = 79\cdot4$. In order to use formula (1) or (2), the value of d is required, and when the calculation of moments has to be made about the centroid vertical its value is, as we have seen above, the same as ν_1'; in the present case it is the first moment about the vertical through age 77. The powers of d are next calculated by logarithms. As it happens d is a comparatively simple number; if it had been ·48327, say, the propriety of using logarithms would have been more obvious

$$d^2 = \cdot2304 \qquad d^3 = + \cdot110592 \qquad d^4 = \cdot0530842$$

Remembering that ν_1 is zero and ν_0' and ν_0 are each unity because they are the total frequency, we reach

$$\nu_2 = 3\cdot2336 \qquad \nu_3 = -1\cdot430976 \qquad \nu_4 = 30\cdot416289$$

It is wise to work to a large number of decimal places because, owing to the subtractions involved, calculations which began with, say, seven figures may end with only five. It is well, therefore, to use a seven-place logarithm table (e.g. Chambers's) and antilogarithm table (e.g. Filipowski's) or a multiplying machine.

It will be noticed that the terms required in the calculation of successive moments can be formed continuously. Thus, in formula (1), we require to calculate the following multiples of ν_2':

1   for the second moment

$3d$,,   third   ,,

$\dfrac{4 \cdot 3}{2} d^2$,,   fourth   ,,

I have found it convenient to adopt a regular system in calculating moments, as in other statistical work, and create the habit of putting results and calculations in fixed positions, so that the arithmetic, which is sometimes complicated, can be followed quickly and can be confirmed or rectified more easily.

9. Although the above is the usual way to calculate moments, another method was suggested by the late Sir G. F. Hardy and used by him in his graduation of the British Offices Tables 1863–93. He pointed out that by summing the statistical numbers and forming a new series in the same way as is done by actuaries when the N_x column is formed from the D_x column and then summing these results (cf. the S column), and so on, equations can be formed which give the same results as the method of moments. The arrangement in Table III shows both the method of calculation and the form of the expression obtained by the process.

Considering the line opposite the first term, we notice that the sum of the series is given, and that the second summation, which we will call S_2 when the total frequency is taken as unity, gives the first moment of the whole distribution about a vertical situated at unit distance before the point corresponding to $f(1)$. Still considering only the first line, we see that S_3 gives each function multiplied by coefficients of the form $\dfrac{n(n+1)}{2!}$ or $\dfrac{n^2+n}{2!}$, i.e. it gives $\dfrac{\nu_2' + \nu_1'}{2}$, where ν' is written for the moment, because by definition the tth moment (ν_t') of the whole distribution is given by the sum of $n^t f(n)$ for all values of n. S_4 and S_5 give each function multiplied by $\dfrac{n^3 + 3n^2 + 2n}{6}$ and $\dfrac{n^4 + 6n^3 + 11n^2 + 6n}{24}$ respectively.

The following equations result:

$$S_2 = \nu_1' \qquad\qquad S_4 = \tfrac{1}{6}(\nu_3' + 3\nu_2' + 2\nu_1')$$
$$S_3 = \tfrac{1}{2}(\nu_2' + \nu_1') \qquad S_5 = \tfrac{1}{24}(\nu_4' + 6\nu_3' + 11\nu_2' + 6\nu_1')$$

These equations enable us to calculate the moments about the selected origin, but if it is necessary to find moments about the mean, the following relations are more convenient; they can be reached by substituting in the above the values in formula (2), and remembering that $S_2 = d$.

$$\nu_2 = 2S_3 - d(1+d)$$
$$\nu_3 = 6S_4 - 3\nu_2(1+d) - d(1+d)(2+d)$$
$$\nu_4 = 24S_5 - 2\nu_3\{2(1+d)+1\} - \nu_2\{6(1+d)(2+d)-1\}$$
$$- d(1+d)(2+d)(3+d)$$

10. Table IV shows the working in the numerical example already dealt with by the direct method. The fifth sum is unnecessary, as the total of the items in the fourth sum gives the only value required:

TABLE IV

Frequency	First sum	Second sum	Third sum	Fourth sum
29	1,000	5,480	19,372	54,508
23	971	4,480	13,892	35,136
81	948	3,509	9,412	21,244
151	867	2,561	5,903	11,832
192	716	1,694	3,342	5,929
239	524	978	1,648	2,587
157	285	454	670	939
93	128	169	216	269
29	35	41	47	53
6	6	6	6	6
Total (for check) 1,000	5,480	19,372	54,508	132,503

From the totals of the columns we have

$$S_2 = d = 5\cdot48 \quad S_3 = 19\cdot372 \quad S_4 = 54\cdot508 \quad \text{and} \quad S_5 = 132\cdot503$$

The first value S_2 or d shows that the mean is at age $52 + 5\cdot48 \times 5 = 79\cdot4$. The age 52 is used because it is the centre of the group before that in which numbers occur, and, as has been already remarked, the summation method assumes the work to be done with reference to this position. The application of the formula for ν_2, ν_3 and ν_4, given above, enables us to find

$$\nu_2 = 3\cdot2336 \quad \nu_3 = -1\cdot43099 \quad \nu_4 = 30\cdot4164$$

11. We may save arithmetical work in several ways when using the summation method. If, instead of making all the calculations implied in Table III we stop at the sums next above the lines ruled in the various columns we shall have as the final totals

$$\Sigma f(n); \quad \Sigma n f(n); \quad \Sigma \frac{n(n-1)}{2!} f(n); \quad \Sigma \frac{n(n-1)(n-2)}{3!} f(n)$$

and

$$\Sigma \frac{n(n-1)(n-2)(n-3)}{4!} f(n).$$

TABLE III.

No. of term	Function	Sum of function	Second sum of function	Third sum of function
1	$f(1)$	$f(1)+f(2)+\ldots+f(n)$	$f(1)+2f(2)+\ldots+nf(n)$	$f(1)+3f(2)+6f(3)+\ldots+\frac{n(n+1}{2}$
2	$f(2)$	$f(2)+f(3)+\ldots+f(n)$	$f(2)+2f(3)+\ldots+(n-1)f(n)$	$f(2)+3f(3)+\ldots+\frac{(n-1)n}{2}f($
3	$f(3)$	$f(3)+\ldots+f(n)$	$f(3)+2f(4)+\ldots+(n-2)f(n)$	$f(3)+3f(4)+\ldots+\frac{(n-2)(n-1}{2}$
4	$f(4)$	$f(4)+\ldots+f(n)$	$f(4)+2f(5)+\ldots+(n-3)f(n)$	$f(4)+3f(5)+\ldots+\frac{(n-3)(n-2)}{2}$
.	.	.	.	.
.	.	.	.	.
.	.	.	.	.
$n-2$	$f(n-2)$	$f(n-2)+f(n-1)+f(n)$	$f(n-2)+2f(n-1)+3f(n)$	$f(n-2)+3f(n-1)+6f(n)$
$n-1$	$f(n-1)$	$f(n-1)+f(n)$	$f(n-1)+2f(n)$	$f(n-1)+3f(n)$
n	$f(n)$	$f(n)$	$f(n)$	$f(n)$
General form of the sums on line 1		$\sum_{1}^{n} f(n)$ $=\nu'_0$	$\sum_{1}^{n} nf(n)$ $=\nu'_1$	$\sum_{n=1}^{n} \frac{n^2+n}{2}f(n)$ $=\frac{1}{2}(\nu'_2+\nu'_1)$

21 a

TABLE III. (*continued*)

Fourth sum of function	Fifth sum of function
$f(1)+4f(2)+\ldots+\dfrac{n(n+1)(n+2)}{3!}f(n)$	$f(1)+5f(2)+\ldots+\dfrac{n(n+1)(n+2)(n+3)}{4!}f(n)$
$f(2)+4f(3)+\ldots+\dfrac{(n-1)n(n+1)}{3!}f(n)$	$f(2)+5f(3)+\ldots+\dfrac{(n-1)n(n+1)(n+2)}{4!}(fn)$
$f(3)+4f(4)+\ldots+\dfrac{(n-2)(n-1)n}{3!}f(n)$	$f(3)+5f(4)+\ldots+\dfrac{(n-2)(n-1)n(n+1)}{4!}f(n)$
$f(4)+4f(5)+\ldots+\dfrac{(n-3)(n-2)(n-1)}{3!}f(n)$	$f(4)+5f(5)+\ldots+\dfrac{(n-3)(n-2)(n-1)n}{4!}f(n)$
$\vdots$	$\vdots$
$f(n-2)+4f(n-1)+10f(n)$	$f(n-2)+5f(n-1)+15f(n)$
$f(n-1)+4f(n)$	$f(n-1)+5f(n)$
$f(n)$	$f(n)$
$\displaystyle\sum_{n=1}^{n}\dfrac{n^3+3n^2+2n}{6}f(n)$ $=\tfrac{1}{6}(v'_3+3v'_2+2v'_1)$	$\displaystyle\sum_{n=1}^{n}\dfrac{n^4+6n^3+11n^2+6n}{24}f(n)$ $=\tfrac{1}{24}(v'_4+6v'_3+11v'_2+6v'_1)$

The formulae of § 9 for S_3, S_4 and S_5 in terms of ν_1', ν_2' etc. require modification only by altering the alternate signs from $+$ to $-$. The form of moment given in this paragraph (§ 11) has been called "factorial moments".* A little further saving of work can be effected by taking the figures up to the totals next below the lines ruled in Table III. This gives the same result as that just obtained if the origin is assumed to be shifted one space. It will suffice if we take this second case for a numerical example and using the figures from Table IV, we should work only so far as 4480 for the second sum, 9412 for the third, 11,832 for the fourth and for the fifth we sum the last six entries in the fourth column and obtain 9783.

12. These are the direct ways of using the summation method, but, as in the multiplication method of calculating moments, we can shorten the work by using a central term instead of the first term as the starting point or arbitrary origin.† We shall now use this arrangement with the "factorial moment" form. A little care is necessary, because, though there is no difficulty about the interpretation of the sums on the positive side of the selected point, the moments for the terms on the negative side assume that multiplications are made by the powers of negative quantities. Table IV (A) gives an example of summations that have to be made. The figure 978 is $\Sigma nf(n)$ for values on the positive side of the arbitrary origin and 498 is the similar sum on the negative side, ignoring sign, say $\Sigma mf(-m)$. The mean is found by dividing the difference, $\Sigma nf(n) - \Sigma mf(-m)$, by the total frequency, i.e. $(978 - 498)/1000 = \cdot 48$. Taking, now, the final figures in the columns headed "Third sum", 670 represents

---

* The semi-invariants (or half-invariants) used by Thiele and other writers can be obtained from moments. The second and third semi-invariants are the same as the second and third moments about the mean and the fourth semi-invariant is the fourth moment less three times the square of the second moment $(\mu_4 - 3\mu_2^2)$.

† I have to thank Mr G. J. Lidstone for the suggestion of shortened summations.

$\Sigma \dfrac{n(n-1)}{2!} f(n)$ and 324 is the corresponding figure on the negative side; similarly with the other columns.

Now reverting to the expressions in §11, which relate only to positive summations, we can write

$$\nu_2' = 2S_3' + S_2'$$
$$\nu_3' = 6S_4' + 6S_3' + S_2'$$
$$\nu_4' = 24S_5' + 36S_4' + 14S_3' + S_2'$$

where S' is used instead of S to indicate the different system of summation.

In Table IV (A) however we have divided the distribution into two parts, and in applying the expressions just given we must work out each part separately, adding the items for even moments and subtracting for odd moments. Hence for the whole distribution

$$\nu_2' = (2 \times \cdot 670 + \cdot 978) + (2 \times \cdot 324 + \cdot 498) = 3 \cdot 464$$
$$\nu_3' = (6 \times \cdot 269 + 6 \times \cdot 670 + \cdot 978) - (6 \times \cdot 139 + 6 \times \cdot 324 + \cdot 498)$$
$$= 3 \cdot 336$$
$$\nu_4' = (24 \times \cdot 059 + 36 \times \cdot 269 + 14 \times \cdot 670 + \cdot 978)$$
$$+ (24 \times \cdot 029 + 36 \times \cdot 139 + 14 \times \cdot 324 + \cdot 498) = 32 \cdot 194$$

and, transferring to the mean,

$$\nu_2 = 3 \cdot 2336, \quad \nu_3 = -1 \cdot 43098 \quad \text{and} \quad \nu_4 = 30 \cdot 41626.$$

We may now express the work in symbols. Writing P for summations on the positive side and N for those on the negative side of the arbitrary origin, we have

$$\nu_1' = P_2 - N_2$$
$$\nu_2' = (2P_3 + P_2) + (2N_3 + N_2)$$
$$= 2(P_3 + N_3) + (P_2 + N_2)$$

and similarly

$$\nu_3' = 6(P_4 - N_4) + 6(P_3 - N_3) + (P_2 - N_2)$$
$$\nu_4' = 24(P_5 + N_5) + 36(P_4 + N_4) + 14(P_3 + N_3) + (P_2 + N_2).$$

TABLE IV (A)

Frequency	First sum	Second sum	Third sum	Fourth sum	Fifth sum
29	29	29	29	29	29
23	52	81	110	139	...
81	133	214	324	...	...
151	284	498	...	...	...
192					
239	524	978	...	...	...
157	285	454	670	...	...
93	128	169	216	269	...
29	35	41	47	53	59
6	6	6	6	6	6
1,000					

13. A comparison of Table IV (A) with Table IV will show that a saving of numerical work is effected by using a central point as the starting point for the summation, for the sums are numerically smaller and the value of S_2 or d, which enters into the formulae on p. 20, is much smaller. It will be readily appreciated that whenever there is a large number of terms the summation method, and especially the form of it given in Table IV (A), is an improvement on the product method of calculating moments. By means of an adding machine the summations can be obtained mechanically with little trouble, even for series containing as many as a hundred terms.

14. In § 12 of Chapter II the mean was described alternatively in terms of the individual observations, $o_1, o_2, \ldots o_N$. Similarly the tth moment is

$$\frac{1}{N} \sum_{i=1}^{N} o_i^t$$

and the tth factorial moment is

$$\frac{1}{N} \sum_{i=1}^{N} o_i(o_i - 1) \ldots (o_i - t + 1) = \frac{1}{N} \sum_{i=1}^{N} o_i^{(t)}$$

15. It is now necessary to consider the calculation of moments from the curve, for until this has been done it is impossible to form equations for finding the constants.

(24)

Let $y_x = f(x, a, b, c, \ldots)$, where $a, b, c, \ldots$ are constants to be determined.

We have seen, on pp. 13 and 14, that one way of working would be to find

$$f(1, a, b, c, \ldots) \times 1^n + f(2, a, b, c, \ldots) \times 2^n + \ldots$$

say,
$$\sum_{x=1}^{r} f(x, a, b, c, \ldots) \times x^n$$

and this would give a result which might be used in forming equations if it were not for the fact that it is often impossible to find an algebraic expression for the sum of such a series in terms of the constants. It is, however, generally possible to find such an expression for the integral, and as we have defined the nth moment of an ordinate y_x as $y_x x^n$, the nth moment of the whole distribution from $x = h$ to $x = k$ is

$$\int_h^k y_x x^n \, dx \quad \text{or} \quad \int_h^k f(x, a, b, c, \ldots) x^n \, dx$$

The total frequency (i.e. total number of cases investigated) is $\int_h^k y_x dx$, and the mean is $\int_h^k y_x x \, dx \Big/ \int_h^k y_x dx$, as we have already noticed.

16. If the moments from the equation to the curve are calculated in this way and equated to the moments calculated from statistics by assuming that the latter consist of a series of ordinates, an inaccuracy is introduced.

Let us consider the two cases:

(1) When the statistics are a system of isolated terms or ordinates* and we wish to pass a curve very closely through them.

(2) When they are a system of areas but the moments are calculated by assuming the areas to be concentrated at the middle points of the bases.

---

* Strictly speaking, not a frequency distribution but a series of values requiring graduation. Distributions have generally to be dealt with as areas for frequency-curve work because they tell the way the whole number of cases is divided in groups, and the whole area between the curve and the axis of x must therefore be used.

17. In case (1) above, the terms $y_0, y_1, y_2, \ldots y_{n-1}$ are given by the statistics, and since $\int_{-\frac{1}{2}}^{\frac{1}{2}} y_x dx$ is approximately equal to y_0, it is simplest* to assume that $\int_{-\frac{1}{2}}^{n-\frac{1}{2}} y_x dx$ is given by the equation to the curve, and we have to find adjustments to counteract the error† caused by equating $\sum_{x=0}^{n-1} Xy_x$ to $\int_{-\frac{1}{2}}^{n-\frac{1}{2}} Xy_x dx$. The most practical way of overcoming the difficulty is by calculating the true area corresponding to the ordinates $y_0, y_1, \ldots y_{n-1}$ by means of a quadrature formula (formula of approximate summation). Many formulae are well known, but for the present purpose it is convenient to have expressions which give approximate values of an area in terms of ordinates lying both within and without the base on which the area to be valued stands. Symbolically, these formulae express $\int_{-\frac{1}{2}}^{\frac{1}{2}} y_x dx$ in terms of $y_{-\frac{1}{2}}, y_{\frac{1}{2}}, y_{-1\frac{1}{2}}, y_{1\frac{1}{2}}$, etc., or $y_0, y_1, y_{-1}, y_2, y_{-2}$, etc.

I. Let
$$y_x = a + bx + cx^2 + dx^3 + ex^4$$

then
$$\int_{-\frac{1}{2}}^{\frac{1}{2}} y_x dx = a + \frac{c}{12} + \frac{e}{80}$$

and
$$y_0 = a$$
$$y_{-1} + y_1 = 2(a + c + e)$$
$$y_{-2} + y_2 = 2(a + 4c + 16e)$$

Now, assume the required integral can be equated to
$$hy_0 + k(y_{-1} + y_1) + l(y_{-2} + y_2)$$
substitute the values given just above and equate the coef-

---

* It is generally possible to use these limits in case (1), but if other limits have to be taken, such as 0 to n, different quadrature formulae must be used.

† Actuarial readers will notice that the error is analogous to that introduced by assuming $(1 + i)^{\frac{1}{2}} a_{\overline{n}|} = \ddot{a}_{\overline{n}|}$.

(26)

ficients of a, c and e respectively to 1, $\frac{1}{12}$ and $\frac{1}{80}$, and we have

$$h + 2k + 2l = 1$$
$$2k + 8l = \tfrac{1}{12}$$
$$2k + 32l = \tfrac{1}{80}$$

The solution of these equations gives

$$h = \tfrac{5178}{5760}, \quad k = \tfrac{308}{5760} \quad \text{and} \quad l = -\tfrac{17}{5760}$$

and we obtain

$$\int_{-\frac{1}{2}}^{\frac{1}{2}} y_x dx = \tfrac{1}{5760}\{5178 y_0 + 308(y_{-1} + y_1) - 17(y_{-2} + y_2)\} \quad \ldots\ldots(\text{I})$$

II. If

$$y_x = a + bx + cx^2 + dx^3$$
$$\int_{-\frac{1}{2}}^{\frac{1}{2}} y_x dx = \tfrac{1}{24}\{y_{-1} + 22 y_0 + y_1\} \qquad\qquad \ldots\ldots(\text{II})$$

III. If

$$y_x = a + bx + cx^2 + dx^3 + ex^4$$
$$\int_{-\frac{1}{2}}^{\frac{1}{2}} y_x dx = \tfrac{1}{1440}\{802(y_{\frac{1}{2}} + y_{-\frac{1}{2}}) - 93(y_{1\frac{1}{2}} + y_{-1\frac{1}{2}}) + 11(y_{2\frac{1}{2}} + y_{-2\frac{1}{2}})\}$$
$$\ldots\ldots(\text{III})$$

IV. If

$$y_x = a + bx + cx^2 + dx^3$$
$$\int_{-\frac{1}{2}}^{1\frac{1}{2}} y_x dx = \tfrac{1}{24}\{27 y_0 + 17 y_1 + 5 y_2 - y_3\} \qquad\qquad \ldots\ldots(\text{IV})$$

18. We can now take the calculation of the moments, where $\int_{-\frac{1}{2}}^{n-\frac{1}{2}} y_x dx$ is required in terms of y_0, y_1, $\ldots$ y_{n-1}.

Now

$$\int_{-\frac{1}{2}}^{n-\frac{1}{2}} y_x dx = \int_{-\frac{1}{2}}^{\frac{1}{2}} y_x dx + \int_{\frac{1}{2}}^{1\frac{1}{2}} y_x dx + \ldots + \int_{n-1\frac{1}{2}}^{n-\frac{1}{2}} y_x dx$$

If formula (I) be applied it can be used for all the integrals on the right-hand side of this equation except the first two and the last two, and the values of these are given by (IV).

(27)

Summing the values obtained and writing (IV) with the denominator 5760, we obtain

$$\int_{-\frac{1}{2}}^{n-\frac{1}{2}} y_x \, dx = \tfrac{1}{5760}\{6463y_0 + 4371y_1 + 6669y_2 + 5537y_3$$
$$+ 5760(y_4 + y_5 + \ldots + y_{n-6} + y_{n-5}) + 5537y_{n-4}$$
$$+ 6669y_{n-3} + 4371y_{n-2} + 6463y_{n-1}\} \quad \ldots\ldots(V)$$

which means that we can multiply

the first and last ordinates by $\tfrac{6463}{5760}(= 1 \cdot 1220486)$,
the second and last but one by $\tfrac{4371}{5760}(= \cdot 7588542)$,
the third and last but two by $\tfrac{6669}{5760}(= 1 \cdot 1578125)$,
the fourth and last but three by $\tfrac{5537}{5760}(= \cdot 9612847)$,

leave all the other ordinates unaltered, and work out the moments in the usual way from this modified series of ordinates. If there are less than eight ordinates, another formula must be evolved.

19. In the following table the original series and the modified one are set out in the first two columns, and in the other columns the calculations of the first four moments about the middle of the range by the direct method are shown:

TABLE V

y_x	Modified by formula (V) y_x'	x	$y_x' \times x$	$y_x' \times x^2$	$y_x' \times x^3$	$y_x' \times x^4$
51·81	58·13	-4	232·52	930·08	3,720·32	14,881·28
43·74	33·19	-3	99·57	288·71	866·13	2,598·39
35·42	41·01	-2	82·02	164·04	328·08	656·16
27·80	26·72	-1	26·72	26·72	26·72	26·72
20·42	20·42		$-440·83$	...	$-4,941·25$	...
13·79	13·26	$+1$	13·26	13·26	13·26	13·26
8·22	9·52	$+2$	19·04	38·08	76·16	152·32
4·29	3·26	$+3$	9·78	29·34	88·02	264·06
1·69	1·90	$+4$	7·60	30·40	121·60	486·40
207·18	207·41		$+ \ 49·68$ $-391·15$	1,520·63	$+ \ \ 299·04$ $-4,642·21$	19,078·59

207·41 is then treated as the total frequency, and the moments for unit frequency (μ_n') would be obtained by dividing

$-391 \cdot 15$, $1520 \cdot 63$, etc. by $207 \cdot 41$, and not by $207 \cdot 18$, which is not the "total frequency", but merely gives the uncorrected sum of certain equidistant values.

20. The work can sometimes be simplified considerably, for if the values at the ends of the experience are very small and have a tendency to keep close to the axis of x before they finally vanish (i.e. if there is high contact; most actuarial functions l_x, a_x, D_x, etc. have high contact at the old age end of the table), then it is reasonable to suppose that ordinates before the first and after the last exist, but are insignificant in value. Thus the integral corresponding to the whole series of ordinates can be legitimately extended beyond the limits $-\frac{1}{2}$ and $n-\frac{1}{2}$ previously used, because the additional area thus introduced will be evanescent. Now if the area be so extended, the effect will be that in equation (V) the significant ordinates from y_0 to y_{n-1} will all have the coefficient unity, and the ordinates with weighted coefficients will all vanish.

The practical result is, that if there is high contact at one end of the statistics the adjustment need only be made at the other end, while if there is high contact at both ends no adjustment is necessary.

Mathematically, high contact means that the first few differential coefficients vanish at the point of contact. The diagrams on pp. 71 and 83 show high contact at both ends of the curves, and the diagram on p. 63 shows high contact at the longer durations.

21. The second case in § 16, namely, that in which mid-ordinates are used instead of areas, may now be examined. By concentrating areas about the middle points of their bases, we assume that the distances by which the areas $\int_{-\frac{1}{2}}^{\frac{1}{2}} y_x dx$, $\int_{\frac{1}{2}}^{1\frac{1}{2}} y_x dx$, etc. must be multiplied are the same as the distances from y_0, y_1, etc.; that is, the tth moment from the statistics is

$$\int_{-\frac{1}{2}}^{+\frac{1}{2}} y_x dx X^t + \int_{\frac{1}{2}}^{1\frac{1}{2}} y_x dx (X+1)^t + \ldots + \int_{n-1\frac{1}{2}}^{n-\frac{1}{2}} y_x dx (X+n-1)^t$$

(29)

and we require $\int_{-\frac{1}{2}}^{n-\frac{1}{2}} (X+x)^t y_x dx$, where X is the distance of y_0 from the ordinate about which moments are calculated.

Applying formula (I) to each integral and collecting terms, we reach as a general coefficient

$$\tfrac{1}{5760}\{\ldots + [5178h^t + 308\{(h-1)^t + (h+1)^t\} - 17\{(h-2)^t + (h+2)^t\}]y + \ldots\}$$

where h is written for $X+x$ for simplification, or

$$\tfrac{1}{5760}\{5760h^t + 240t(t-1)h^{t-2} + 3t(t-1)(t-2)(t-3)h^{t-4} + \ldots\}$$

If $t = 1$ this becomes h

,, $t = 2$	,,	$h^2 + \tfrac{1}{12}$
,, $t = 3$	,,	$h^3 + \tfrac{1}{4}h$
,, $t = 4$	,,	$h^4 + \tfrac{1}{2}h^2 + \tfrac{1}{80}$

It has already been noticed that if there is high contact, the value of $\int_{-\frac{1}{2}}^{n-\frac{1}{2}} (X+x)^t y dx$ is found by using the unadjusted ordinates; that is, the second moment is given by a series, the general term of which is h^2y; the third by a series, the general term of which is h^3y, and so on; hence, if μ be written for the true adjusted moment about the mean and ν for the unadjusted moment, the relations between μ and ν are given by

$$\mu_2 + \tfrac{1}{12} = \nu_2 \quad \text{or} \quad \mu_2 = \nu_2 - \tfrac{1}{12}$$
$$\mu_4 + \tfrac{1}{2}\mu_2 + \tfrac{1}{80} = \nu_4 \quad \text{or} \quad \mu_4 = \nu_4 - \tfrac{1}{2}\nu_2 + \tfrac{7}{240}$$

The mean needs no adjustment, for if $t = 1$ the general term has the correct coefficient h, and the third moment has to be adjusted by $\tfrac{1}{4}$ of the first moment, which is zero where the moments are taken about the mean.* In order to demonstrate the correction for the nth moment by the above method, a parabola of at least the nth order is necessary. If we apply these adjustments to the moments found on p. 19, for Example IV of Table I, we have $\mu_2 = 3 \cdot 1503$, $\mu_3 = -1 \cdot 430976$ and

* These adjustments were first given by W. F. Sheppard in *Proc. Lond. Math. Soc.* XXIX, 353–80. See also Appendix I.

$\mu_4 = 28 \cdot 82866$. These adjustments are found to make an appreciable difference in the constants obtained from the moments, especially when there is a small number of terms.

In other words, they allow for the grouping, and the lesson to be learnt is that a moderate amount of grouping saves work and, thanks to our knowledge of the correct adjustments, does not introduce error in the circumstances described.

22. The practical conclusions in the two preceding paragraphs as to the treatment of moments when there is high contact can be checked numerically. The equation to a curve with high contact at each end having been written down, we can work out the ordinates at equidistant points or the areas on equal bases and calculate the moments from the figures. From the equation to the curve we can also calculate the area and moments for the whole curve and it will be found that the corresponding figures agree. A good curve with which to make experiments in this way is "the normal curve of error" because the ordinates and areas are accurately tabulated in *Tables for Statisticians*, but anyone wishing to apply this sort of check is advised to wait until he has read a little about frequency-curves.

When there is not high contact at both ends of the curve, the adjustments become more difficult to value; suggestions have been made for finding the corrections, and this matter is further discussed in Appendix I, but a beginner is advised to avoid these refinements.

A student should calculate the moments for one or two distributions, and make the easier adjustments; he can also find the standard deviations of distributions, for the S.D. $= \sqrt{\mu_2}$, where the μ_2 has been adjusted in accordance with the above rules. In Examples III and IV of Table I there is clearly high contact, and in Example I the rough moment should be used. In Examples II and V there is more doubt, and in the calculation of the moments for Example II (see p. 60), no adjustment was made.

This advice is given not because adjustment is unnecessary,

but because a beginner can content himself with mastering the general idea and leave out some of the refinements until he has a little more experience. Later on, when the methods of Appendix I are examined, it will be seen that Sheppard's adjustments alone do not usually improve the rough moments when the distribution is abrupt.

23. Before proceeding to deal with fitting more complicated curves it is advisable to consider the application of the method of moments to a simple case, namely, when

$$y = a + bx + cx^2 + \dots$$

Let the range be $2l$, and let the origin be at the middle point of the range, and m_0 stand for the area and m_n for the nth moment of the whole distribution about the middle of the range. Then

$$m_{2s} = \int_{-l}^{+l} (a + bx + cx^2 + \dots) x^{2s} \, dx$$

$$= 2l \times l^{2s} \left(\frac{a}{2s+1} + \frac{cl^2}{2s+3} + \dots \right)$$

and similarly $m_{2s+1} = 2l \times l^{2s+1} \left(\dfrac{bl}{2s+3} + \dfrac{dl^3}{2s+5} + \dots \right)$

These equations show that the even moments give the constants a, c, e, etc., and the odd moments give the constants b, d, f, etc. This is, of course, the result of using moments about the middle of the range, and makes the solution of the equations less laborious than they would otherwise have been. The solution can also be simplified a little by writing

$$\frac{1}{2l} \cdot \frac{m_{2s}}{l^{2s}} = \frac{a}{2s+1} + \frac{cl^2}{2s+3} + \dots$$

so that

$$\left.\begin{aligned}
\frac{1}{2l} \cdot m_0 &= a + \frac{cl^2}{3} + \frac{el^4}{5} + \dots \\[2mm]
\frac{1}{2l} \cdot \frac{m_2}{l^2} &= \frac{a}{3} + \frac{cl^2}{5} + \frac{el^4}{7} + \dots \\[2mm]
\frac{1}{2l} \cdot \frac{m_4}{l^4} &= \frac{a}{5} + \frac{cl^2}{7} + \frac{el^4}{9} + \dots
\end{aligned}\right\}$$

and similarly

$$
\left.
\begin{aligned}
\frac{1}{2l}\cdot\frac{m_1}{l} &= \frac{bl}{3}+\frac{dl^3}{5}+\frac{fl^5}{7}+\ldots \\[2mm]
\frac{1}{2l}\cdot\frac{m_3}{l^3} &= \frac{bl}{5}+\frac{dl^3}{7}+\frac{fl^5}{9}+\ldots \\[2mm]
\frac{1}{2l}\cdot\frac{m_5}{l^5} &= \frac{bl}{7}+\frac{dl^3}{9}+\frac{fl^5}{11}+\ldots
\end{aligned}
\right\}
$$

The solution of these equations gives the constants required, for example,

(i) if $y = a + bx$, we have

$$a = \frac{1}{2l}m_0$$

$$b = \frac{3}{l}\cdot\frac{1}{2l}\cdot\frac{m_1}{l}$$

(ii) if $y = a + bx + cx^2$

$$a = \frac{3}{4}\left\{\frac{3}{2l}\cdot m_0 - \frac{5}{2l}\cdot\frac{m_2}{l^2}\right\}$$

$$b = \frac{3}{l}\cdot\frac{1}{2l}\cdot\frac{m_1}{l}$$

$$c = \frac{15}{4l^2}\left\{-\frac{1}{2l}\cdot m_0 + \frac{3}{2l}\cdot\frac{m_2}{l^2}\right\}$$

(iii) if $y = a + bx + cx^2 + dx^3$

$$a = \frac{3}{4}\left\{\frac{3}{2l}\cdot m_0 - \frac{5}{2l}\cdot\frac{m_2}{l^2}\right\}$$

$$b = \frac{15}{4l}\left\{\frac{5}{2l}\cdot\frac{m_1}{l} - \frac{7}{2l}\cdot\frac{m_3}{l^3}\right\}$$

$$c = \frac{15}{4l^2}\left\{-\frac{1}{2l}\cdot m_0 + \frac{3}{2l}\cdot\frac{m_2}{l^2}\right\}$$

$$d = \frac{35}{4l^3}\left\{-\frac{3}{2l}\cdot\frac{m_1}{l} + \frac{5}{2l}\cdot\frac{m_3}{l^3}\right\}$$

The above results, which can easily be extended if it is wished, may now be applied to one or two numerical examples.

24. As a first example, we shall graduate the statistics in Table V, § 19, for which the moments about the middle of the range have been calculated. Taking the curve $y = a + bx + cx^2$, the following values from Table V will be required:

$$2l = 9 \quad \text{or} \quad l = 4 \cdot 5$$
$$m_0 = \quad\ 207 \cdot 41$$
$$m_1 = -\ 391 \cdot 15$$
$$m_2 = \quad 1520 \cdot 63$$

Hence
$$a = \frac{3}{4}\left\{\frac{622 \cdot 23}{9} - \frac{5}{9} \times \frac{1520 \cdot 63}{(4 \cdot 5)^2}\right\}$$
$$= 20 \cdot 563$$

$$b = \frac{3}{4 \cdot 5} \times \frac{1}{9} \times \frac{-391 \cdot 15}{4 \cdot 5}$$
$$= -6 \cdot 4387$$

$$c = \frac{15}{4(4 \cdot 5)^2}\left\{-\frac{207 \cdot 41}{9} + \frac{3}{9} \times \frac{1520 \cdot 63}{(4 \cdot 5)^2}\right\}$$
$$= \cdot 36815$$

25. The best way to obtain the ordinates corresponding to this graduation is by calculating $b + c$ the first difference, and $2c$ the second difference, from the middle term; their values are $-6 \cdot 0706$ and $\cdot 7363$ respectively. Since second differences are constant, the work is done continuously, and is as follows:

	$\varDelta$	$\varDelta^2$
52·208	− 9·016	·736
43·192	− 8·279	
34·913	− 7·543	
27·370	− 6·807	
20·563	− 6·071	
14·492	− 5·335	
9·157	− 4·599	
4·558	− 3·862	
·696		

These graduated figures will be found to agree fairly well with those given in the first column of Table V.

26. As a further example the following statistics, taken from a paper by S. H. J. W. Allin (*J. Inst. Actu.* XXXIX, 350), and giving the values of annuities to widows in pension funds according to the age of the member, may be considered:

Age	Value of annuity	Modified by formula (V) p. 28 a'	Distance from middle of range multiplied by 2 d	$a' \times d$	$a' \times d^2$	$a' \times d^3$
27	21·20	23·79	− 7	166·53	1165·71	8159·97
32	19·91	15·11	− 5	75·55	377·75	1888·75
37	19·34	22·40	− 3	67·20	201·60	604·80
42	18·58	17·86	− 1	17·86	17·86	17·86
				− 327·14		− 10671·38
47	16·74	16·09	+ 1	16·09	16·09	16·09
52	15·69	18·17	+ 3	54·51	163·53	490·59
57	14·70	11·15	+ 5	55·75	278·75	1393·75
62	12·99	14·58	+ 7	102·06	714·42	5000·94
		139·15		+ 228·41	2935·71	+ 6901·37
				− 98·73		− 3770·01

In calculating the above moments it has been assumed that the figures to be graduated represent a system of ordinates; if they had represented a system of areas, the adjustment by formula (V) would have been unsuitable.

The alternative is to avoid the integral calculus and work out from the equation $y = f(x)$ the sum of the ordinates and the moments of the ordinates. In the particular case where $f(x) = a + bx + cx^2 + \dots$ this is practicable, but there are many expressions which, with their moments, can be integrated but do not lend themselves to finite summation. We have therefore confined attention to the more general method.

When there is an even number of terms the difficulty of calculating the moments about the middle of the range is that the terms have to be multiplied by ·5, 1·5, 2·5, etc., and if the series to be graduated contains only a few terms, it is best to deal with the distance d, in the way shown above, and then divide the totals by 2, 4 and 8, in order to obtain the first, second and third moments respectively. In this way, we have

$$l = \quad 4$$
$$m_0 = \quad 139\cdot15$$
$$m_1 = - \quad 49\cdot36$$
$$m_2 = \quad 733\cdot93$$
$$m_3 = -471\cdot25$$

We will now fit the statistics with each of the three curves, the formulae for which have been given, and compare the resulting graduations.

(i) $y = 17\cdot394 - 1\cdot157x$

(ii) $y = 17\cdot633 - 1\cdot157x - \cdot0451x^2$

(iii) $y = 17\cdot633 - 1\cdot190x - \cdot0451x^2 + \cdot0035x^3$

The following table shows the graduations:

Age	Ungraduated	(i)	(ii)	(iii)
27	21·20	21·44	21·13	21·13
32	19·91	20·29	20·24	20·28
37	19·34	19·13	19·27	19·31
42	18·58	17·97	18·20	18·22
47	16·74	16·82	17·04	17·02
52	15·69	15·66	15·80	15·76
57	14·70	14·50	14·46	14·43
62	12·99	13·34	13·03	13·05

Formulae (ii) and (iii) are practically identical, and both are considerably closer to the original figures than (i).

27. The results obtained so far may be summarised as follows:

(1) The method of moments is a general method of finding the constants in a formula suitable to a particular statistical example, and it consists of equating the values of $\Sigma f(n) \times n^t$ (which is called the tth moment, and is

summed for all values of n that occur) to similar expressions obtained from the graduation formula. These latter expressions will be algebraic, and simultaneous equations have to be solved in order to find the arithmetical constants.

(2) The moments from the statistics can be calculated by multiplying the frequencies by appropriate values of n^t or by the summation method.

(3) If moments have been obtained about any one vertical, they can be transferred to any other by the formulae in § 6 of this chapter.

(4) Since the moments from the graduation formula must generally be found by means of the integral calculus, while those from the statistics are found by summation, the latter have to be adjusted before the equations for obtaining the constants can be correctly formed. The adjustments depend on whether the statistics are a system of ordinates or a system of areas; in the former case adjustment is made by equation (V), and in the latter by the formulae in § 21 if there is high contact at both ends of the curve.

CHAPTER IV

PEARSON'S SYSTEM OF FREQUENCY-CURVES

1. When it becomes necessary in practical work to decide on a system of curves for describing frequency distributions, we have to bear in mind that

(1) Any expression used must be a graduation formula; it must remove the roughness of the material.

(2) There must not be so many constants in the formula that we require a great number of moments, for this means that the accuracy is reduced. The higher the moment the more liable it is to error when deduced from ungraduated observations; this is clear, when we remember that the ends of the experiences are multiplied by the highest numbers and their powers.

(3) There must be a systematic method of approaching frequency distributions.

2. Now, considering the more obvious characteristics of frequency distributions, we find they generally start at zero, rise to a maximum, and then fall sometimes at the same but often at a different rate. At the ends of the distribution there is often high contact. This means, mathematically, that a series of equations $y = f(x)$, $y = \phi(x)$, etc. must be chosen, so that in each equation of the series $dy/dx = 0$ for certain values of x, namely, at the maximum and at the end of the curve where there is contact with the axis of x.

(38)

The above suggests that dy/dx may be put equal to $\dfrac{y \times (x+a)}{F(x)}$;
then, if $y = 0$, $dy/dx = 0$, and there is, therefore, contact at
one end of the curve, while if $x = -a$, $dy/dx = 0$, and we
have the maximum we require. So long as $F(x)$ is general the
form assumed for dy/dx is extremely general and includes
cases when dy/dx may not be zero when y is zero. If $F(x)$ is
expanded by Maclaurin's theorem in ascending powers of x,
we have

$$\frac{dy}{dx} = \frac{y(x+a)}{b_0 + b_1 x + b_2 x^2 + \ldots} \qquad \ldots\ldots(\text{I})$$

We shall return to this equation and show how it can be put in
the form $y = f(x)$, so as to express y as a direct function of x,
and we shall see that we have obtained something more general
than is implied at the beginning of this paragraph. We shall
obtain curves taking various widely different shapes. As the
matter has up to the present been approached from an experi-
mental point of view, it will be interesting to see how equation
(I) can be obtained up to the x^2 term in the denominator from
elementary propositions in the theory of probabilities.

3. If p be the probability of an event happening and q the
probability of its failing, then the probabilities of its happening
once, twice, and so on out of n trials are given by the terms of
the expansion $(p+q)^n$; or if we have N cases, the terms of
$N(p+q)^n$ give the frequency distribution of the N cases into
$n+1$ groups. The binomial series does not represent nearly all
the probabilities that arise, and another series that occurs is the
hypergeometrical. Thus the chances of getting r, $r-1$, $\ldots$, 0
black balls from a bag containing pn black and qn white balls
when r balls are drawn, are given by the successive terms of the
series

$$\frac{pn(pn-1)\ldots(pn-r+1)}{n(n-1)\ldots(n-r+1)}$$
$$\times \left\{ 1 + \frac{rqn}{pn-r+1} + \frac{r(r-1)}{2!}\frac{qn(qn-1)}{(pn-r+1)(pn-r+2)} + \ldots \right\}$$

A numerical example may help to make clear the way the series arises. A bag contains seven balls, of which four are black and three white; then if three balls are drawn the probability that

$$\text{All will be black is} \quad \frac{4 \cdot 3 \cdot 2}{7 \cdot 6 \cdot 5}$$

$$\text{Two will be black is} \quad \frac{4 \cdot 3 \cdot 3}{7 \cdot 6 \cdot 5} \times {}_3C_1$$

$$\text{One will be black is} \quad \frac{4 \cdot 3 \cdot 2}{7 \cdot 6 \cdot 5} \times {}_3C_2$$

$$\text{None will be black is} \quad \frac{3 \cdot 2 \cdot 1}{7 \cdot 6 \cdot 5}$$

The sum of these four expressions is unity. The terms can be seen to agree with the series by putting $n = 7$, $pn = 4$, $qn = 3$, and $r = 3$.

Other series may arise, but those given will be sufficient for the present purpose, and we shall proceed to consider how they can be put in the form of equation (I). The inconvenience of the expressions as they now stand becomes fairly obvious when an attempt is made to calculate numerical values for a large number of groups, and besides this, they are not continuous, while the statistics of practical work often are.

Considering the hypergeometrical series, and remembering that the function required for equation (I) is $\dfrac{1}{y}\dfrac{dy}{dx}$, and that, as the series is discontinuous, finite differences must be used, we have

$$y_x = \frac{pn(pn-1)\ldots(pn-r+1)}{n(n-1)\ldots(n-r+1)} \cdot \frac{r(r-1)\ldots(r-x+2)}{(x-1)!}$$
$$\times \frac{qn(qn-1)\ldots(qn-x+2)}{(pn-r+1)(pn-r+2)\ldots(pn-r+x-1)}$$

$$\Delta y_x = y_{x+1} - y_x = y_x \left\{ \frac{r-x+1}{x} \cdot \frac{qn-x+1}{pn-r+x} - 1 \right\}$$
$$= y_x \left\{ \frac{(r+1)(qn+1)-x(n+2)}{x(pn-r+x)} \right\} \quad \text{for} \quad p+q=1$$

(40)

and

$$y_{x+\frac{1}{2}} = \tfrac{1}{2}(y_{x+1} + y_x)$$

$$= \tfrac{1}{2}y_x \left\{ \frac{(r+1)(qn+1) - x[2(r+1) + n(q-p)] + 2x^2}{x(pn-r+x)} \right\}$$

$$\therefore \quad \frac{\Delta y_x}{y_{x+\frac{1}{2}}} = \frac{2\{(r+1)(qn+1) - x(n+2)\}}{(r+1)(qn+1) - x\{2(r+1) + n(q-p)\} + 2x^2}$$

which may be put in the form of equation (I),

$$\frac{1}{y}\frac{dy}{dx} = \frac{a+x}{b_0 + b_1 x + b_2 x^2}$$

In this form the actuarial reader will naturally think of the force of mortality: to proceed from the force of mortality, after changing its sign, to the "number living" (l_x) in a life table is the same thing as to proceed from the formula just given to a frequency-curve.

4. Returning to equation (I), we see that it can be written in the form

$$(b_0 + b_1 x + b_2 x^2 + \ldots)\frac{dy}{dx} = y(x+a)$$

multiplying each side by x^n, and integrating with respect to x, we have

$$\int x^n (b_0 + b_1 x + b_2 x^2 + \ldots)\frac{dy}{dx}dx = \int y(x+a)x^n dx$$

Integrate the left-hand side by parts treating dy/dx as one part, and the right-hand side as the sum of two functions, and then

$$x^n(b_0 + b_1 x + b_2 x^2 + \ldots)y - \int \{nb_0 x^{n-1} + (n+1)b_1 x^n$$
$$+ (n+2)b_2 x^{n+1} + \ldots\}y\,dx$$
$$= \int yx^{n+1}dx + \int yax^n dx$$

or, if at the ends of the range of the curve the expression $x^n(b_0 + b_1 x + b_2 x^2 + \ldots)y$ vanishes, we have

$$-nb_0\mu'_{n-1} - (n+1)b_1\mu'_n - (n+2)b_2\mu'_{n+1} - \ldots = \mu'_{n+1} + a\mu'_n$$

where we use the notation we have already adopted, namely,

$$\mu_n' = \int y x^n dx$$

If we put $n = 0, 1, 2, \ldots s$ respectively, we get $s + 1$ equations to enable us to find a, b_0, b_1, ... etc., in terms of the moments (μ') as shown by the following equations, which have been obtained by writing the equation in the form

$$a\mu_n' + nb_0\mu_{n-1}' + (n + 1) b_1\mu_n' + (n + 2) b_2\mu_{n+1}' + \ldots = -\mu_{n+1}'$$

and then putting $n = 0, 1, 2$, etc.

$$\left.\begin{aligned}
a\mu_0' + 0 \times b_0 + b_1\mu_0' + 2b_2\mu_1' + \ldots &= -\mu_1' \\
a\mu_1' + b_0\mu_0' + 2b_1\mu_1' + 3b_2\mu_2' + \ldots &= -\mu_2' \\
a\mu_2' + 2b_0\mu_1' + 3b_1\mu_2' + 4b_2\mu_3' + \ldots &= -\mu_3' \\
a\mu_3' + 3b_0\mu_2' + 4b_1\mu_3' + 5b_2\mu_4' + \ldots &= -\mu_4'
\end{aligned}\right\} \quad \ldots\ldots\text{(II)}$$

etc., etc.

Let us now make $\mu_1' = 0$, and alter the other moments in the way indicated in Chapter III, for the result of making $\mu_1' = 0$ is to change the origin of the system to the mean of the distribution. We can also treat μ_0' as 1, and these simplifications lead to the following results:

(1) Keeping b_0 only, we have

$$\frac{1}{y}\frac{dy}{dx} = -\frac{x}{\mu_2}$$

(2) Keeping b_0 and b_1, the first three equations in the system (II) above give

$$a + b_1 = 0$$

$$b_0 = -\mu_2$$

and

$$a\mu_2 + 3b_1\mu_2 = -\mu_3$$

or

$$b_1 = -\frac{\mu_3}{2\mu_2}$$

and

$$a = \frac{\mu_3}{2\mu_2}$$

(42)

and the differential equation becomes

$$\frac{1}{y}\frac{dy}{dx} = -\frac{x + \dfrac{\mu_3}{2\mu_2}}{\mu_2 + \dfrac{\mu_3}{2\mu_2}x}$$

(3) Keeping b_0, b_1, b_2, the system gives

$$a + b_1 = 0$$

$$b_0 + 3b_2\mu_2 = -\mu_2$$

$$a\mu_2 + 3b_1\mu_2 + 4b_2\mu_3 = -\mu_3$$

$$a\mu_3 + 3b_0\mu_2 + 4b_1\mu_3 + 5b_2\mu_4 = -\mu_4$$

The solution of these simultaneous equations is perfectly straightforward, and leads to

$$\frac{1}{y}\frac{dy}{dx} = -\frac{x + \dfrac{\mu_3(\mu_4 + 3\mu_2{}^2)}{10\mu_2\mu_4 - 18\mu_2{}^3 - 12\mu_3{}^2}}{\dfrac{\mu_2(4\mu_2\mu_4 - 3\mu_3{}^2)}{10\mu_2\mu_4 - 18\mu_2{}^3 - 12\mu_3{}^2} + \dfrac{\mu_3(\mu_4 + 3\mu_2{}^2)}{10\mu_2\mu_4 - 18\mu_2{}^3 - 12\mu_3{}^2}x + \dfrac{2\mu_2\mu_4 - 3\mu_3{}^2 - 6\mu_2{}^3}{10\mu_2\mu_4 - 18\mu_2{}^3 - 12\mu_3{}^2}x^2}$$

In this last form put $\beta_1 = \dfrac{\mu_3^2}{\mu_2^3}$ and $\beta_2 = \dfrac{\mu_4}{\mu_2^2}$ and

$$\frac{1}{y}\frac{dy}{dx} = -\frac{x + \dfrac{\sqrt{\mu_2}\,\sqrt{\beta_1}(\beta_2 + 3)}{2(5\beta_2 - 6\beta_1 - 9)}}{\dfrac{\mu_2(4\beta_2 - 3\beta_1) + \sqrt{\mu_2}\,\sqrt{\beta_1}(\beta_2 + 3)\,x + (2\beta_2 - 3\beta_1 - 6)\,x^2}{2(5\beta_2 - 6\beta_1 - 9)}}$$

$$\cdots\cdots(\text{III})$$

5. The reasoning by which equation (I) was first obtained showed that a is the distance between the origin and the mode, or as the origin has now been transferred to the mean by putting $\mu_1' = 0$, a is the distance between the mean and the mode. This distance in terms of the moments is, therefore,

$$\frac{\sigma\sqrt{\beta_1}(\beta_2 + 3)}{2(5\beta_2 - 6\beta_1 - 9)}$$

where σ is the standard deviation $\sqrt{\mu_2}$.

Since the skewness is the distance between the mean and mode divided by the standard deviation

$$\text{Skewness} = \frac{\sqrt{\beta_1}(\beta_2 + 3)}{2(5\beta_2 - 6\beta_1 - 9)}$$

6. It would be possible to obtain constants in the differential equation (I) by using a greater number of terms and retaining b_3, b_4, etc., but there are strong practical objections to such a course. Besides the increase in arithmetical work, the gain in introducing additional constants is small because the higher moments become untrustworthy, as we have already noticed. Karl Pearson has shown* that "we might easily on a random sample reach a 7th or 8th moment having half or double the value it actually has in the general population. Constants based on these high moments will be practically idle. They may enable us to describe closely an individual random sample, but no safe argument can be drawn from this individual sample as to the general population at large, at any rate so far as the argument is based on the constants depending on these high moments." In some actuarial statistics where there are as many as 100,000 cases, it might be worth while to go as far as the next term of the series, but even here the value of the work is discounted because any other smaller body of statistics on the same subject could not be compared satisfactorily with the result. For practical purposes it is probable that the equation taken as far as b_2 will be sufficient, and we shall confine our attention to the forms thus obtained.

7. Turning to the particular form of equation (I) given in equation (III) it will be seen that it is possible to obtain a formula representing the statistics by inserting in that equation the values of the moments found from the statistics, but this would not give a graduation in the same form as that in which the original data appeared, for in the latter we have y, while the former gives $\dfrac{1}{y}\dfrac{dy}{dx}$ or $\dfrac{d\log y}{dx}$. It would, therefore,

* "Skew correlation and non-linear regression", *Drapers' Company Res. Mem.* 1905, p. 9. See also Chapter X.

be necessary to integrate the expression we obtain in order to get terms comparable with the original data, and it is better in practical work to deal with the equations in the forms in which we require them for comparison, rather than by using the differential equation and then integrating the result. The latter method could only give proportional not actual frequencies.

8. The next step is, therefore, to replace the equation

$$\frac{d \log y}{dx} = \frac{x+a}{b_0 + b_1 x + b_2 x^2}$$

by one of the form $y = f(x)$, and to do this $\dfrac{x+a}{b_0 + b_1 x + b_2 x^2}$ must be integrated.

Let us consider equation (III) as a general expression for integration, then we notice that the form the integral takes depends on the particular values of the coefficients of x in the denominator. The problem is, in fact, merely a consideration of the forms taken by the denominator for

$$b_0 + b_1 x + b_2 x^2$$
$$= b_2 \left[x - \frac{-b_1 + \sqrt{(b_1^2 - 4b_0 b_2)}}{2b_2} \right] \left[x - \frac{-b_1 - \sqrt{(b_1^2 - 4b_0 b_2)}}{2b_2} \right]$$

and the criterion for fixing the form in a particular case is, obviously, the same as that for the nature of the roots of the equation $b_0 + b_1 x + b_2 x^2 = 0$, viz. $b_1^2/(4b_0 b_2)$, which, by substituting from formula (III), gives

$$\frac{\beta_1(\beta_2 + 3)^2}{4(2\beta_2 - 3\beta_1 - 6)(4\beta_2 - 3\beta_1)} \qquad \ldots\ldots(IV)$$

9. If expression (IV) is negative the roots are real and of different sign, and we get one of the main types of curve—called Type I by Karl Pearson, to whom this system of curves is due; if expression (IV) is positive and less than unity the roots are complex, and we get the second main type (Pearson's Type IV), and if expression (IV) is positive and greater than

unity the roots are real and of the same sign, and we reach the third main type (Pearson's Type VI).

This really covers the whole field, but in the limiting cases when one type changes into another we reach simpler forms of transition curves. Thus when the criterion is large (theoretically infinite) one root is ∞ (Type III), when it is unity the two roots are equal (Type V), and when it is zero the roots are equal in magnitude but of opposite sign (Type II). If in the last case $b_1 = b_2 = 0$, we reach what we shall call the "normal curve of error": this name is open to some objection just as are the other names given to it (e.g. Probability curve, Gaussian curve, etc.). Then again the expression for $(d \log y)/dx$ may be reducible to the form $a'/(b_0' + b_1'x)$ and we have a binomial or a straight line for the frequency-curve (cf. Types VIII, IX and XI), while if the expression reduces to a constant the curve is the ordinary geometrical progression which we are pleased to find as a special case of a system of frequency-curves because we are already familiar with it in the theory of probability in connection with sequences from coin tossing, etc. As we proceed we shall find that in certain circumstances the curves may be **J**-shaped or even **U**-shaped, with limits of a single ordinate or two separated ordinates. A diagram at the end of the book will give the reader an idea of the variety of shapes taken by the curves evolved from the formula

$$\frac{d \log y}{dx} = \frac{x + a}{b_0 + b_1 x + b_2 x^2}$$

In practice we shall require the equations to the various kinds of frequency-curve, and we shall also want to know which type should be used in a particular case. We cannot usually guess the type from the appearance of the rough data and need an arithmetical test.

10. We will deal first with the equations to the frequency-curves, that is, with the actual integration, and begin with the three main types.

First Main Type (*Pearson's Type I*). The factors in the

(46)

denominator, when the roots of $b_0 + b_1 x + b_2 x^2 = 0$ are real and of different signs, take the form

$$b_2 \left[x - \frac{-b_1 + \sqrt{\text{a positive quantity}}}{2b_2} \right]$$

$$\times \left[x - \frac{-b_1 - \sqrt{\text{a positive quantity}}}{2b_2} \right]$$

and the expression to be integrated is therefore of the form

$$\frac{1}{b_2} \cdot \frac{x+a}{(x+A_1)(x-A_2)} = \frac{1}{b_2} \cdot \frac{A_1 - a}{A_1 + A_2} \cdot \frac{1}{x+A_1} + \frac{1}{b_2} \cdot \frac{A_2 + a}{A_1 + A_2} \cdot \frac{1}{x-A_2}$$

by partial fractions.

The integration is now simple, and gives

$$\log y = \frac{1}{b_2} \cdot \frac{A_1 - a}{A_1 + A_2} \log(x + A_1) + \frac{1}{b_2} \cdot \frac{A_2 + a}{A_1 + A_2} \log(x - A_2)$$
$$+ \text{a constant}$$

$$\therefore \qquad y = y'(x + A_1)^{\frac{1}{b_2} \cdot \frac{A_1 - a}{A_1 + A_2}} (x - A_2)^{\frac{1}{b_2} \cdot \frac{A_2 + a}{A_1 + A_2}}$$

where y' results from the constant introduced by integration.

If the origin is now transferred to the mode (i.e. put x for $x + a$), we have

$$y = y_0 \left(1 + \frac{x}{a_1}\right)^{m_1} \left(1 - \frac{x}{a_2}\right)^{m_2}$$

where $m_1/a_1 = m_2/a_2$.

Second Main Type (Pearson's Type IV). If the roots of the equation $b_0 + b_1 x + b_2 x^2 = 0$ are complex, it is impossible to throw the denominator into real factors; and when this occurs we have to integrate by putting the expression on the right-hand side of the fundamental differential equation in the form

$$\frac{X + c}{b_2(X^2 + A^2)}$$

where $X = x + \dfrac{b_1}{2b_2}$, $c = a - \dfrac{b_1}{2b_2}$ and $A^2 = \dfrac{b_0}{b_2} - \dfrac{b_1^2}{4b_2^2}$

(47)

Then $\log y = \int \dfrac{X+c}{b_2(X^2+A^2)}\,dX$

$= \int \dfrac{X}{b_2(X^2+A^2)}\,dX + \int \dfrac{c}{b_2(X^2+A^2)}\,dX$

$= \dfrac{1}{2b_2}\log(X^2+A^2) + \dfrac{c}{Ab_2}\tan^{-1}\dfrac{X}{A} + \text{constant}$

$\therefore \quad y = y'(X^2+A^2)]^{1/2b_2}\,e^{c/Ab_2\,\tan^{-1}X/A}$

or $\quad y = \dfrac{y_0}{\left(1+\dfrac{x^2}{a^2}\right)^m}\,e^{-\nu\tan^{-1}x/a}$

where a has a meaning different from that implied in equation (I). The relation between this type and Type I can be seen by factorising the denominator of the right-hand side of the differential equation, $b_2(X-iA)(x+iA)$, and then obtaining an expression for y having the same form as Type I, but containing complex expressions.

Third Main Type (Pearson's Type VI). The factorising is the same as Type I, but the roots of the equation being of like sign, the factors of the denominator take the form $(x+A_1)(x+A_2)$. The work is then the same, but at the end the origin is put by Pearson not at the mode but so that one of the expressions $x+A_1$ or $x+A_2$ can be written as x. The form is then

$$y = y_0(x-a)^{m_1}x^{-m_2}$$

11. We may now set out a few of the transition types.

Pearson's *Type II* is the same as his Type I when $a_1 = a_2$.

Type III. This type is reached when the criterion is ∞, which happens when $b_2 = 0$.

$\log y = \int \dfrac{x+a}{b_0+b_1 x}\,dx$

$= \int\left(\dfrac{1}{b_1} + \dfrac{a-b_0/b_1}{b_1 x+b_0}\right)dx$

$= \dfrac{x}{b_1} + \left(a-\dfrac{b_0}{b_1}\right)\dfrac{1}{b_1}\log(b_1 x+b_0) + \text{constant}$

(48)

and
$$y = y' e^{x/b_1} (b_1 x + b_0)^{(a - b_0/b_1)/b_1}$$

or, by changing the origin,

$$y = y_0 e^{-\gamma x} \left(1 + \frac{x}{a} \right)^{\gamma a}$$

where a has a meaning different from that implied in equation (I). This type can be seen to be a particular case of Type I when a_2 becomes infinite.

Type V. In this case, when the roots are real and equal,

$$
\begin{aligned}
\log y &= \int \frac{1}{b_2} \frac{x + a}{(x + b_1/2b_2)^2} dx \\
&= \int \frac{1}{b_2} \frac{(x + b_1/2b_2) + (a - b_1/2b_2)}{(x + b_1/2b_2)^2} dx \\
&= \int \frac{dx}{b_2(x + b_1/2b_2)} + \int \frac{a - b_1/2b_2}{b_2(x + b_1/2b_2)^2} dx \\
&= \frac{1}{b_2} \log(x + b_1/2b_2) - \frac{a - b_1/2b_2}{b_2(x + b_1/2b_2)} + \text{constant}
\end{aligned}
$$

$$\therefore \qquad y = y'(x + b_1/2b_2)^{\frac{1}{b_2}} e^{-\frac{a - b_1/2b_2}{b_2(x + b_1/2b_2)}}$$

$$= y_0 x^{-p} e^{-\gamma/x}$$

Normal Curve of Error. Putting

$$b_1 = b_2 = 0$$

$$
\begin{aligned}
\log y &= \int \frac{x + a}{b_0} dx \\
&= \frac{x^2}{2b_0} + \frac{ax}{b_0} + \text{constant} \\
&= \frac{(x + a)^2}{2b_0} + \text{constant}
\end{aligned}
$$

$$\therefore \qquad y = y' e^{(x + a)^2/2b_0}$$

or, by changing the origin and altering the constant,

$$y = y_0 e^{-x^2/c}$$

In a similar way the other less important transition curves can be obtained. These are

$$\left(1+\frac{x}{a}\right)^{-m}; \quad \left(1+\frac{x}{a}\right)^{m}; \quad e^{-x/\sigma}; \quad x^{-m}$$

and we reach J-shaped curves when in Type I either m_1 or m_2 is negative and U-shaped curves when both are negative. J-shaped curves can also be obtained with Type III.

12. A table is inserted which gives the list of curves with Pearson's numbering and with the origin as he generally uses it. This is convenient because in reading other work on the subject it will be found that Pearson's numbering, etc. is usually adopted. I have, however, added a note of the equation to each curve when the origin is at the mean. There is something to be said for uniformity as regards the origin, and the mean is convenient because all distributions have means and the moments are worked out about the vertical through the mean. A column in the table gives criteria to show which curve should be used in an individual case.

We may here deal with a little difficulty that students sometimes encounter in connection with types which may be expressed in the same algebraic form (e.g. Types VIII, IX and XI can all be written hx^k). The question may be asked why we should not fit hx^k from a to b and find h, k, a and b from the equations for the moments. The answer is that the criteria afford in effect a simplification of the equations and automatically tell us a good deal about the value of the constants and the range of the curve.

13. We shall return to some of the technical points when discussing numerical examples in the next chapter but may now recapitulate the method, and see the steps that have to be taken to fit a frequency-curve to statistics.

(1) Arrange the statistics in sequence.

(2) Calculate the moments about a convenient vertical.

(3) Transfer the moments to the centroid vertical (vertical through the mean).

(4) If there is high contact at both ends of the curve, apply Sheppard's adjustments to the moments (i.e. deduct $\frac{1}{12}$ and $\frac{1}{2}\nu_2 - \frac{7}{240}$ from the second and fourth moments respectively). If there is not high contact, see Appendix I.

(5) Calculate the criterion.

(6) By means of Table VI decide which curve should be used.

As an alternative to (5) and (6), the curve to be used can be found from diagrams in *Tables for Statisticians*, which show the type in terms of β_1 and β_2.

TABLE VI. *Frequency-curves*

No. of type usually adopted (Pearson)	Equation to curve in form usually adopted (Pearson)		Equation with origin at mean
	Equation	Origin	
MAIN TYPES			
I	$y = y_0(1 + x/a_1)^{\nu a_1}(1 - x/a_2)^{\nu a_2}$	Mode	$y = y_e(1 + x/A_1)^{m_1}(1 - x/A_2)^{m_2}$ where $(m_1 + 1)/A_1 = (m_2 + 1)/A_2$
IV	$y = y_0(1 + x^2/a^2)^{-m}e^{-\nu\tan^{-1}x/a}$	$\nu a/(2m - 2)$ after mean	$y = y_0[1 + (x/a - \nu/r)^2]^{-m}e^{-\nu\tan^{-1}(x/d -}$ where $r = 2m - 2$
VI	$y = y_0(x - a)^{q_2}x^{-q_1}$	a before start of curve	$y = y_e(1 + x/A_1)^{-q_1}(1 + x/A_2)^{q_2}$ where $(q_1 - 1)/A_1 = (q_2 + 1)/A_2$
TRANSITION TYPES			
"Normal curve"	$y = y_0 e^{-x^2/2\sigma^2}$	Mode (= mean)	$y = y_0 e^{-x^2/2\sigma^2}$
II	$y = y_0(1 - x^2/a^2)^m$	Mode (= mean)	$y = y_0(1 - x^2/a^2)^m$
VII	$y = y_0(1 + x^2/a^2)^{-m}$	Mode (= mean)	$y = y_0(1 + x^2/a^2)^{-m}$
III	$y = y_0(1 + x/a)^{\gamma a}e^{-\gamma x}$	Mode	$y = y_e(1 + x/A)^p e^{-\gamma x}$ where $A = (p + 1)/\gamma$ and $p = \gamma a$
V	$y = y_0 x^{-p}e^{-\gamma/x}$	Start of curve	$y = y_e(1 + x/A)^{-p}e^{(p-2)/(1+x/A)}$
VIII	$y = y_0(1 + x/a)^{-m}$	End of curve	$y = y_e(1 + x/A)^{-m}$
IX	$y = y_0(1 + x/a)^m$	End of curve	$y = y_e(1 + x/A)^m$
X	$y = y_0 e^{-x/\sigma}$	Start of curve	$y = y_e e^{-x/\sigma}$
XI	$y = y_0 x^{-m}$	b before start	$y = y_e(1 + x/A)^{-m}$
XII	$y = y_0\left(\dfrac{\sigma\{\sqrt{(3 + \beta_1)} + \sqrt{\beta_1}\} + x}{\sigma\{\sqrt{(3 + \beta_1)} - \sqrt{\beta_1}\} - x}\right)^{\sqrt{(\beta_1/3+\beta_1)}}$	Mean	$y = y_e(1 + x/A_1)^{m_1}(1 - x/A_2)^{m_2}$ is an alternative form where m_1 and are equal numerically and < 1 but opposite sign

51a

TABLE VI. *Frequency-curves* (continued)

Criterion	Remarks	For calculation of constants see p.
κ negative	Limited range; skew; usually bell-shaped, but may be U-shaped, J-shaped or twisted J-shaped	58
$\kappa > 0$ and < 1	Unlimited range; skew; bell-shaped	66
$\kappa > 1$	Unlimited range in one direction; skew; bell-shaped, but may be J-shaped	76
$\kappa = 0$, $\beta_1 = 0$, $\beta_2 = 3$	Unlimited range; symmetrical; bell-shaped	80
$\kappa = 0$, $\beta_1 = 0$, $\beta_2 < 3$	Limited range; symmetrical; usually bell-shaped, but U-shaped when $\beta_2 < 1 \cdot 8$	86
$\kappa = 0$, $\beta_1 = 0$, $\beta_2 > 3$	Unlimited range; symmetrical; bell-shaped	90
$2\beta_2 = 6 + 3\beta_1$	Unlimited range in one direction; usually bell-shaped, but may be J-shaped	92
$\kappa = 1$	Unlimited range in one direction; bell-shaped	96
κ negative; $\lambda = 0$; $5\beta_2 - 6\beta_1 - 9$ negative	Range from infinite ordinate at $-a$ to finite ordinate at 0 (or from $-a(1-m)/(2-m)$ to $a/(2-m)$ with origin at mean)	102
κ negative; $\lambda = 0$; $5\beta_2 - 6\beta_1 - 9$ positive, $2\beta_2 - 3\beta_1 - 6$ negative	Range from $x = -a$ where $y = 0$ to $x = 0$ where $y = y_0$ (or from $-a(m+1)/(m+2)$ to $a/(m+2)$ with origin at mean)	105
$\beta_1 = 4$, $\beta_2 = 9$	Exponential from finite ordinate at 0 (or $-\sigma$ with origin at mean) to infinitesimal ordinate at ∞; J-shaped	108
$\kappa > 1$, $\lambda = 0$, $2\beta_2 - 3\beta_1 - 6$ positive	J-shaped; starts at $x = b$ (or $-b/(m-2)$ with origin at mean) where ordinate is finite	110
$5\beta_2 - 6\beta_1 - 9 = 0$	Twisted J-shaped; special case of Type I	111

$$\kappa = \frac{\beta_1(\beta_2 + 3)^2}{4(4\beta_2 - 3\beta_1)(2\beta_2 - 3\beta_1 - 6)}, \quad \beta_1 = \mu_3^2/\mu_2^3, \quad \beta_2 = \mu_4/\mu_2^2, \quad \lambda = \frac{(4\beta_2 - 3\beta_1)(10\beta_2 - 12\beta_1 - 18)^2 - \beta_1(\beta_2 + 3)^2(8\beta_2 - 9\beta_1 - 12)}{(3\beta_1 - 2\beta_2 + 6)\{\beta_1(\beta_2 + 3)^2 + 4(4\beta_2 - 3\beta_1)(3\beta_1 - 2\beta_2 + 6)\}}$$

51b

CHAPTER V

CALCULATION

1. The next point to be considered is the calculation of the constants for any particular distribution, when the moments have been calculated and the type to be used has been decided. The formulae required for the numerical work will be given for each type; a numerical example, including the calculation of the graduated figures, will follow, with the proofs of the formulae.

2. Some general points relating to the calculation of the curves when the constants have been found may be conveniently considered here. When the constants are known, we can calculate the ordinate for any value of x by substituting that value in the expression for the frequency-curve; and if areas are required, some method of proceeding from ordinates to areas must be found. The most simple is probably to calculate mid-ordinates, and then by the quadrature formula (I) or (II) on p. 27 find the areas. It is occasionally more convenient to calculate the ordinates at the beginning of each group, and then formula (III) should be used. These formulae can be best applied in the form of differences; thus, from (II) we have

$$\int_{-\frac{1}{2}}^{\frac{1}{2}} y_x dx = y_0 - \tfrac{1}{24}\{\varDelta y_{-1} - \varDelta y_0\}$$

from (I)

$$\int_{-\frac{1}{2}}^{\frac{1}{2}} y_x dx = y_0 - \tfrac{291}{5760}\{\varDelta y_{-1} - \varDelta y_0\} + \tfrac{17}{5760}\{\varDelta y_{-2} - \varDelta y_1\}$$

from (III)

$$\int_{-\frac{1}{2}}^{\frac{1}{2}} y_x dx = \tfrac{1}{2}\{y_{\frac{1}{2}} + y_{-\frac{1}{2}}\} + \tfrac{82}{1440}\{\varDelta y_{-1\frac{1}{2}} - \varDelta y_{\frac{1}{2}}\} - \tfrac{11}{1440}\{\varDelta y_{-2\frac{1}{2}} - \varDelta y_{1\frac{1}{2}}\}$$

Formula (II) is generally sufficiently accurate, while the others will be found to give a result true to five figures in ordinary cases—exceptional cases will be referred to in the numerical examples that follow.

3. It is sometimes a help to see the graduation expressed graphically, and this has been done with some of the examples. The best method is to insert a vertical height y_0 at the mode; note the ends of the curve, and the heights of the ordinates that have been calculated. These heights give points on the curve, which can be drawn through them fairly easily. In drawing the curve, as well as in calculating the constants, the sign of the skewness must be borne in mind, for it is possible to draw the curve with the skewness on the wrong side of the mode, and if the distribution is nearly symmetrical, it is not so easy to notice the mistake as one might expect. The tangent to the curve at the mode is parallel to the axis of x except in the case of the J-shaped curves or some of the less common transition types.

4. It is best to draw on a rather large scale in order to gain distinctness, and the curves given here were drawn larger than their present size; the reduction being, of course, made in the process of reproduction.

The base elements should also be fairly large in proportion to the height, so that the curve may not ascend too steeply; otherwise small horizontal differences between the graduated and ungraduated curves are apt to conceal large vertical differences when the curve is rising or falling rapidly, but it is the latter differences that are of importance.

5. The reader should notice that all the cases considered in the following pages assume complete distributions, and it is in general only possible to find the curve from part of a distribution by means of successive approximation which is extremely laborious. Another point, to which reference will again be made, is with regard to grouping statistics; it is sometimes impossible to obtain many groups, but for accuracy in finding moments the greater the number of groups the better, unless the total number of cases is small. A little discretion is needed in this respect, but in actuarial statistics which are sometimes based on as many as 200,000 cases, seventy groups might be used for great accuracy. In our examples we have grouped merely to save work, space and printing, and the grouping does not alter the method.

If there is high contact so that we know the proper adjustments, grouping leads to little or no error. An adjustment of one-twelfth to the second moment when ten ages are grouped and used as the unit has much more effect, proportionately, than when only five ages are grouped or when individual ages are used. The fear sometimes expressed that grouping destroys accuracy has no proper foundation in such cases; a little numerical evidence on this point will be found in Appendix I.

6. Another matter with which it seems advisable to deal here is connected with the criterion, κ. This may have any value from $-\infty$ to $+\infty$, and from the following diagram it will be seen how the types cover all the possible values of the criterion and do not overlap.

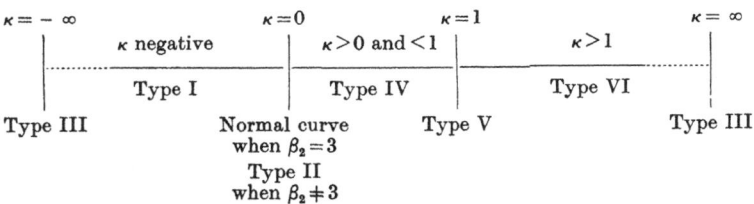

Just before $\kappa = 0$, Type I becomes nearly symmetrical, and after that value is passed we have a skew curve of unlimited range, and so on. At each critical point there are one or more "transition" curves. If by a mistake a student should use the wrong main type, he will find his mistake by reaching an imaginary quantity in one of the square roots which occur in the equations for the constants, but transition types can be used when the values of the criterion approximate to the theoretical values; they can, in fact, be viewed as approximations which give an accurate result in a limiting case. It is impossible to give exact limits within which we are justified in using a transition type; theoretically, as we shall see later, the justification depends on the size of the standard error of the function dealt with, but in practice we can be guided to a great extent by the size of the experience; if there are few cases, a larger deviation in the criterion will arise than if there are many. Individual cases must be considered on their merits, but if the student finds himself in doubt he can avoid using the transition type and be on the safe side in the matter of accuracy. The student has one safe guide in every case, namely, that "the proof of the pudding is in the eating". He should try transition curves in a few cases where he has little hope of their applicability and compare the results with those obtained by the right main types and he will then learn much about both classes of curves.

7. In the formulae that are given for the various types, the choice of sign for a square root depends on the sign of μ_3. If the frequency is concentrated more closely before the mean than after it, the mode is on the left-hand side of the mean and μ_3 is positive; the signs of certain constants in each type must therefore depend on the sign of μ_3 in order that the mode and mean may lie in their correct relative positions. Where, however, no remark is made as to the sign of the expression in which a square root is given the positive root is implied, and the reader will find that these rules become easier to follow when he has worked out two examples, one giving a positive and the other a negative value for μ_3. Thus, if we imagine the frequencies in the example for Type I to be written in the opposite order 1, 3, 7, 13, etc., all the numerical work would be the same, but m_1 would be $2 \cdot 776978$, $m_2 = \cdot 409833$, $a_1 = 13 \cdot 52728$, and $a_2 = 1 \cdot 99638$, and the graduation would be the same, but the numbers in the columns of the table on p. 62 would run in the opposite order.

8. The arithmetical work is heavy and in some respects unfamiliar to most students. There is no royal road to success

(55)

in it except care, system and the use of common-sense at the final stage. It is irritating at the end of a lengthy piece of arithmetic to find a slip at an early stage and to have to re-calculate, but these slips become fewer and of less importance with experience, for when we are in practice we suspect a large error immediately an erroneous value is reached. Personally I use seven-figure logarithms as a rule and put a check on every step, although not necessarily to the last figure. This plan was followed with the arithmetical work in this book. The check might not disclose a slip which did not affect the graduation or only affected the final figures of a constant or coefficient. Thus if the last three figures of $\log y_0$ on p. 61 were wrong (which I have no reason to suppose) the mistake would be regrettable, but the graduation in the table on the following page would be unaffected. Moreover, difficulty may be found in reproducing exactly the numerical result of another calcu-lator, owing to the usual unreliability of the end figures when many operations have been made. In lengthy arithmetic the two final figures may be unreliable and two arithmetical processes may both be correct and yet give divergencies. This does not mean that five-figure logarithms are as good as seven, for if seven figures give five figures accurately, we assume that generally speaking five figure work will only be reliable to three figures.

FORMULAE FOR MOMENTS

THESE FORMULAE APPLY TO ALL THE TYPES OF CURVES

$$\left.\begin{array}{l}\nu_1' = d \\ \nu_2 = \nu_2' - d^2 \\ \nu_3 = \nu_3' - 3d\nu_2 - d^3 \\ \nu_4 = \nu_4' - 4d\nu_3 - 6d^2\nu_2 - d^4\end{array}\right\} \text{ or } \left\{\begin{array}{l}\nu_1' = d \\ \nu_2 = \nu_2' - d^2 \\ \nu_3 = \nu_3' - 3d\nu_2' + 2d^3 \\ \nu_4 = \nu_4' - 4d\nu_3' + 6d^2\nu_2' - 3d^4\end{array}\right.$$

or $S_2 = d$

$$\nu_2 = 2S_3 - d(1+d)$$

$$\nu_3 = 6S_4 - 3\nu_2(1+d) - d(1+d)(2+d)$$

$$\nu_4 = 24S_5 - 2\nu_3\{2(1+d)+1\} - \nu_2\{6(1+d)(2+d)-1\} \\ - d(1+d)(2+d)(3+d)$$

$$\left.\begin{array}{l}\mu_2 = \nu_2 - \frac{1}{12} \\ \mu_3 = \nu_3 \\ \mu_4 = \nu_4 - \frac{1}{2}\nu_2 + \frac{7}{240}\end{array}\right\}$$ Sheppard's adjustments when the curve has high contact at both ends

σ (standard deviation) $= \sqrt{\mu_2}$

$\beta_1 = \mu_3^2/\mu_2^3$

$\beta_2 = \mu_4/\mu_2^2$

$$\kappa = \frac{\beta_1(\beta_2+3)^2}{4(4\beta_2 - 3\beta_1)(2\beta_2 - 3\beta_1 - 6)}$$

FIRST MAIN TYPE (TYPE I)

$$y = y_0 \left(1+\frac{x}{a_1}\right)^{m_1} \left(1-\frac{x}{a_2}\right)^{m_2}$$

where
$$m_1/a_1 = m_2/a_2$$

Origin at mode

The values to be calculated in order are

$$r = 6(\beta_2 - \beta_1 - 1)/(6 + 3\beta_1 - 2\beta_2)$$

$$a_1 + a_2 = \tfrac{1}{2}\sqrt{\mu_2}\sqrt{\{\beta_1(r+2)^2 + 16(r+1)\}}$$

The m's are given by

$$\frac{1}{2}\left\{r - 2 \pm r(r+2)\sqrt{\frac{\beta_1}{\beta_1(r+2)^2 + 16(r+1)}}\right\}$$

when μ_3 is positive m_2 is the positive root

$$y_0 = \frac{N}{a_1 + a_2} \cdot \frac{m_1{}^{m_1} m_2{}^{m_2}}{(m_1 + m_2)^{m_1 + m_2}} \cdot \frac{\Gamma(m_1 + m_2 + 2)}{\Gamma(m_1 + 1)\,\Gamma(m_2 + 1)}$$

$$\text{Mode} = \text{Mean} - \frac{1}{2} \cdot \frac{\mu_3}{\mu_2} \cdot \frac{r+2}{r-2}$$

If expressing curve with origin at mean (see Table VI facing p. 51)

$$A_1 + A_2 = a_1 + a_2$$

$$(m_1 + 1)/A_1 = (m_2 + 1)/A_2$$

$$y_e = \frac{N}{A_1 + A_2} \cdot \frac{(m_1 + 1)^{m_1}(m_2 + 1)^{m_2}}{(m_1 + m_2 + 2)^{m_1 + m_2}} \cdot \frac{\Gamma(m_1 + m_2 + 2)}{\Gamma(m_1 + 1)\,\Gamma(m_2 + 1)}$$

For table of Γ functions see p. 266, or *Tables for Statisticians*.

The usual shape of the curve is like that of the following example, but if m_1 and m_2 are approximately equal it is nearly symmetrical, if m_1 and m_2 are not small it tails off at both ends, and if both m_1 and m_2 are small it rises abruptly at both ends. If m_1 is negative the curve is J-shaped; it starts at an infinite ordinate, falls rapidly and runs out at a fixed point (for numerical example see p. 126). If both m_1 and m_2 are negative, the curve is U-shaped, starting and ending with infinite ordinates and having an anti-mode instead of a mode as the usual origin (for numerical example see p. 112). In the J- and U-shaped curves, though the ordinate is infinite, the area is finite. Care is needed in these cases when taking out the Γ function for $\Gamma(t)$ is required when $t < 1$ and the tables give $\log \Gamma(1 + t)$, i.e., $\log t + \log \Gamma(t)$. In the case of J-shaped curves it is best to use the form with origin at the mean or express the curve in the form $y'x^{m_1}(a_1 + a_2 - x)^{m_2}$ with the origin at the start of the curve and

$$y' = \frac{N}{(a_1 + a_2)^{m_1 + m_2 + 1}} \cdot \frac{\Gamma(m_1 + m_2 + 2)}{\Gamma(m_1 + 1)\,\Gamma(m_2 + 1)}$$

An interesting variant of the J-shaped curve arises when m_1 and m_2 are both arithmetically less than unity and one of them is negative. The shape is then like that of No. (11) in the diagram of curves at the end of the book, i.e. it is of twisted J-shape (for example and further notes, see pp. 111–3).

EXAMPLE

As an example of this type the figures in Table I (Example II) may be used. The moments were first found by the summation method (see Chapter III, § 9) as shown in the following table. The reader can check the result by recalculating the moments by the more direct method, taking age 42 as the arbitrary origin. This is how I should myself usually do the work; I only use the summation method when the series is a very long one, and I give it here merely by way of example.

Central age of group	Exposed to risk Example II of Table I	First sum	Second sum	Third sum	Fourth sum
17	34	1,000	5,175	19,809	64,389
22	145	966	4,175	14,634	44,580
27	156	821	3,209	10,459	29,946
32	145	665	2,388	7,250	19,487
37	123	520	1,723	4,862	12,237
42	103	397	1,203	3,139	7,375
47	86	294	806	1,936	4,236
52	71	208	512	1,130	2,300
57	55	137	304	618	1,170
62	37	82	167	314	552
67	21	45	85	147	238
72	13	24	40	62	91
77	7	11	16	22	29
82	3	4	5	6	7
87	1	1	1	1	1
Totals	1,000	5,175	19,809	64,389	186,638

$$S_2 = 5175/1000 = 5\cdot175$$
$$S_3 = 19809/1000 = 19\cdot809$$
$$S_4 = 64389/1000 = 64\cdot389$$
$$S_5 = 186638/1000 = 186\cdot638$$

The next step is to find the moments about the centroid vertical using the formulae on p. 57, and, in this case, as no adjustments* were made in the moments the ν's and μ's are the same.

$$\mu_2 = 7\cdot66237 \qquad \beta_1 = \cdot5072955$$
$$\mu_3 = 15\cdot1069 \qquad \beta_2 = 2\cdot935110$$
$$\mu_4 = 172\cdot326$$

From the values of β_1 and β_2 the criterion (κ) can be calculated, and its value being $-\cdot2645$ shows that Type I must be used (see Table VI).

* The moments should have been adjusted by one of the methods suitable when the curve is abrupt. These have been discussed since the example was prepared, and it was unnecessary to recalculate—see, however, Appendix I. Similar qualifications apply to a few of the other examples.

$$r = 5\cdot186811 \qquad \log r \quad = \cdot7149004$$
$$r+1 = 6\cdot186811 \qquad \log(r+1) = \cdot7914669$$
$$r+2 = 7\cdot186811 \qquad \log(r+2) = \cdot8565363$$
$$r-2 = 3\cdot186811 \qquad \log(r-2) = \cdot5033563$$

The values of $\log(r+1)$, etc. were checked by a Gauss-logarithm table.

$$a_1 + a_2 = 15\cdot52366 \qquad\qquad a_1 = \quad 1\cdot99638$$
$$m_1 = \quad \cdot409833 \qquad\qquad a_2 = 13\cdot52728$$
$$m_2 = \quad 2\cdot776978 \quad \text{Mean} - \text{mode} = \quad 2\cdot223116$$

It will be noted that the expression $\sqrt{\{\beta_1(r+2)^2 + 16(r+1)\}}$ occurs in both the values of $(a_1 + a_2)$ and m.

The mean is at age $12 + 5\cdot175 \times 5 = 37\cdot8750$, and the mode at age $37\cdot8750 - 2\cdot223116 \times 5 = 26\cdot75942$.

The skewness is $\cdot8032$.

The calculation of $\log y_0$ is as follows:

$$\log N = 3\cdot00000$$
$$\operatorname{colog}(a_1 + a_2) = \overline{2}\cdot80901$$
$$m_1 \log m_1 = \overline{1}\cdot84123$$
$$m_2 \log m_2 = 1\cdot23179$$
$$\operatorname{colog}(r-2)^{r-2} = \overline{2}\cdot39590$$
$$\log \Gamma(r) = 1\cdot50406$$
$$\operatorname{colog} \Gamma(m_1 + 1) = \quad \cdot05219$$
$$\operatorname{colog} \Gamma(m_2 + 1) = \overline{1}\cdot34037$$
$$\overline{\log y_0 = 2\cdot17455}$$

where, of course, $\log \Gamma(m_2 + 1) = \log \Gamma(3\cdot776978) = \log 2\cdot776978 + \log 1\cdot776978 + \log \Gamma(1\cdot776978)$, the last value being taken from the table at the end of the book.

The work to this point gives as the curve for graduating the statistics

$$y = 149\cdot47 \left\{1 + \frac{x}{1\cdot99638}\right\}^{\cdot409833} \left\{1 - \frac{x}{13\cdot52728}\right\}^{2\cdot776978}$$

where the origin is at age $26\cdot75942$ and the unit is five years.

The following table shows the calculation of ordinates of the curve from the equation just given:

Age (1)	$1+\dfrac{x}{a_1}$ (2)	$1-\dfrac{x}{a_2}$ (3)	log (2) (4)	log (3) (5)	$m_1 \times$ col.(4) (6)	$m_2 \times$ col.(5) (7)	col. (6) +col. (7) +log y_0 = log y_x (8)	y_x (9)	(10)
17	·02228	1·14429	$\bar{2}$·34792	0·05854	$\bar{1}$·3229	0·1626	1·6601	45·7	44
22	·52319	1·07037	$\bar{1}$·71866	·02955	1·8847	·0821	2·1404	138·2	137
27	1·02410	·99644	0·01034	$\bar{1}$·99845	0·0042	$\bar{1}$·9957	2·1745	149·5	149
32	1·52501	·92252	·18327	·96498	·0751	·9027	2·1525	142·1	142
37	2·02592	·84859	·30662	·92870	·1257	·8020	2·1023	126·6	127
42	2·52683	·77466	·40257	·88911	·1650	·6921	2·0317	107·6	108
47	3·02774	·70074	·48111	·84556	·1972	·5711	1·9429	87·7	88
52	3·52865	·62681	·54760	·79714	·2244	·4367	1·8357	68·5	69
57	4·02956	·55289	·60526	·74264	·2481	·2853	1·7080	51·0	51
62	4·53047	·47896	·65615	·68030	·2689	·1122	1·5557	36·0	36
67	5·03136	·40504	·70169	·60750	·2876	$\bar{2}$·9100	1·3722	23·6	24
72	5·53229	·33111	·74291	·51997	·3045	·6670	1·1461	14·0	14
77	6·03320	·25719	·78055	·41025	·3199	·3623	·8568	7·2	7
82	6·53411	·18326	·81519	·26307	·3341	$\bar{3}$·9535	·4622	2·9	3
87	7·03502	·10934	·84726	·03878	·3472	·3307	$\bar{1}$·8525	·7	1
92	7·53593	·03541	·87714	$\bar{2}$·54913	·3595	$\bar{5}$·9709	3·5050	...	...

Cols. (2) and (3) have a constant first difference, viz. $1/a_1$ or ·500907, and $1/a_2$ or ·073925. The value at any point having been calculated and checked, the other items are formed continuously. Cols. (4)–(9) explain themselves, but we may remark that it is generally advisable to use a larger number of figures than five in taking logarithms, especially if m_1 or m_2 is large. A little care is necessary in multiplying such numbers as $\bar{1}$·71866 by m_1(·409833). If an arithmometer is used, m_1 is put on the plate, and is multiplied by $-$·28134, and the result $-$·1153 must be put in the form $\bar{1}$·8847, to enable us to add it to other logarithms. Col. (10) gives the area, and was formed by applying one of the formulae on p. 52. The area of the first group must be treated separately, as the curve starts at age 16·7775, and the base of the group is therefore 2·7225 in length, instead of 5 years as in the other cases. A good way to find the area is to calculate the ordinates for the middle and ends of the base, and apply Simpson's rule, viz.:

$$\int_0^1 y_x dx = \tfrac{1}{6}\{y_0 + 4y_{\frac{1}{2}} + y_1\}$$

remembering to multiply the result by $\dfrac{2\cdot7225}{5}$ to allow for the different length of the base.

The mid-ordinate is $92\cdot1$, the ordinate at the end of the base is $116\cdot5$, and the ordinate at the start is of course zero; the area is approximately*

$$\frac{2\cdot7225}{5} \times \tfrac{1}{6}\{0 + 4 \times 92\cdot1 + 116\cdot5\} = 44$$

Some people find it better when calculating the ordinates to use the form given in the Notes on p. 59, with the origin at the start of the curve; it avoids bringing in the reciprocals of a_1 and a_2. The columns of $\log x$ and $\log(a_1 + a_2 - x)$ can be formed continuously with the aid of Gauss-logarithms. The initial values will have to be calculated and as a check one or perhaps two other values.

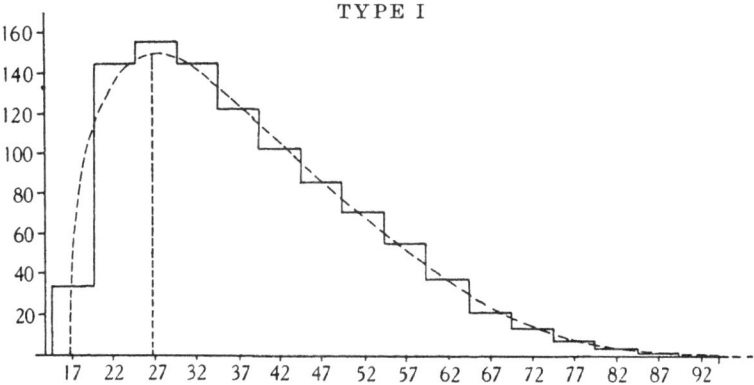

TYPE I

PROOF OF FORMULAE†

The equation to the curve is $y = y_0\left(1 + \dfrac{x}{a_1}\right)^{m_1}\left(1 - \dfrac{x}{a_2}\right)^{m_2}$, where $m_1/a_1 = m_2/a_2$.

* For greater accuracy use more ordinates or Tables of incomplete B-functions.

† The reader who has little acquaintance with formulae of reduction and the Γ and B functions, should consult Appendix II before reading the proofs of the formulae for this and the other types.

Let $a_1 + a_2 = b$ and $z = \dfrac{a_1 + x}{a_1 + a_2}$.

The area from $x = -a_1$ to $x = +a_2$ is the total frequency N.

$$\therefore \quad N = \int_{-a_1}^{a_2} y_0 \left(1 + \frac{x}{a_1}\right)^{m_1} \left(1 - \frac{x}{a_2}\right)^{m_2} dx$$

$$= \int_{-a_1}^{a_2} \frac{y_0}{a_1^{m_1} a_2^{m_2}} (a_1 + x)^{m_1} (a_2 - x)^{m_2} dx$$

$$= \int_0^1 \frac{y_0}{a_1^{m_1} a_2^{m_2}} [z(a_1 + a_2)]^{m_1} [(1 - z)(a_1 + a_2)]^{m_2} (a_1 + a_2) \, dz$$

$$= \int_0^1 \frac{y_0 (a_1 + a_2)^{m_1 + m_2 + 1}}{a_1^{m_1} a_2^{m_2}} z^{m_1} (1 - z)^{m_2} dz$$

$$= \frac{y_0 (m_1 + m_2)^{m_1 + m_2} (a_1 + a_2)}{m_1^{m_1} m_2^{m_2}} B(m_1 + 1, \, m_2 + 1)$$

Or $\quad y_0 = \dfrac{N}{b} \cdot \dfrac{m_1^{m_1} m_2^{m_2}}{(m_1 + m_2)^{m_1 + m_2}} \cdot \dfrac{\Gamma(m_1 + m_2 + 2)}{\Gamma(m_1 + 1)\,\Gamma(m_2 + 1)}$

Using the same method for the moments as that just given for the area, we see that the nth moment, about the line parallel to the axis of y through $x = -a_1$, is

$$N\mu_n' = \int_{-a_1}^{a_2} \frac{y_0}{a_1^{m_1} a_2^{m_2}} (a_1 + x)^n (a_1 + x)^{m_1} (a_2 - x)^{m_2} dx$$

$$= \int_0^1 \frac{y_0 (a_1 + a_2)^{m_1 + m_2 + n + 1}}{a_1^{m_1} a_2^{m_2}} z^{m_1 + n} (1 - z)^{m_2} dz$$

$$= \frac{y_0 (m_1 + m_2)^{m_1 + m_2} b^{n+1}}{m_1^{m_1} m_2^{m_2}} \cdot \frac{\Gamma(m_1 + n + 1)\,\Gamma(m_2 + 1)}{\Gamma(m_1 + m_2 + n + 2)}$$

Now, since $\Gamma(p) = (p - 1)\,\Gamma(p - 1)$, the moments about the line parallel to the axis of y through $x = -a_1$ are as follows:

$$\mu_1' = \frac{b(m_1 + 1)}{m_1 + m_2 + 2}$$

$$\mu_2' = \frac{b^2 (m_1 + 1)(m_1 + 2)}{(m_1 + m_2 + 2)(m_1 + m_2 + 3)} \quad \text{and so on.}$$

Changing the origin in order to get moments about the mean and writing $m_1' = m_1 + 1$ and $m_2' = m_2 + 1$ and $r = m_1' + m_2'$, we have

$$\mu_2 = \frac{b^2 m_1' m_2'}{r^2(r+1)}$$

$$\mu_3 = \frac{2b^3 m_1' m_2'(m_2' - m_1')}{r^3(r+1)(r+2)}$$

$$\mu_4 = \frac{3b^4 m_1' m_2'\{m_1' m_2'(r-6) + 2r^2\}}{r^4(r+1)(r+2)(r+3)}$$

We can simplify these expressions to obtain the equations on p. 58 by writing $\beta_1 = \mu_3^2/\mu_2^3$, $\beta_2 = \mu_4/\mu_2^2$, and $\epsilon = m_1' m_2'$; then

$$\beta_1 = \frac{4(r^2 - 4\epsilon)(r+1)}{\epsilon(r+2)^2} \quad \text{or} \quad \frac{\beta_1(r+2)^2}{4(r+1)} = \frac{r^2}{\epsilon} - 4$$

and

$$\beta_2 = \frac{3(r+1)\{2r^2 + \epsilon(r-6)\}}{\epsilon(r+2)(r+3)}$$

$$\text{or} \quad \frac{\beta_2(r+2)(r+3)}{3(r+1)} = \frac{2r^2}{\epsilon} + r - 6$$

Eliminating r^2/ϵ we find

$$\frac{\beta_1(r+2)^2}{2(r+1)} - \frac{\beta_2(r+2)(r+3)}{3(r+1)} = -8 - r + 6$$

Dividing out by $r + 2$ we have

$$r = \frac{6(\beta_2 - \beta_1 - 1)}{3\beta_1 - 2\beta_2 + 6}$$

From the equation $\dfrac{\beta_1(r+2)^2}{4(r+1)} = \dfrac{r^2}{\epsilon} - 4$ we have

$$\epsilon = \frac{r^2}{4 + \tfrac{1}{4}\beta_1 \dfrac{(r+2)^2}{r+1}}$$

and from the equation for μ_2

$$b^2 = \frac{\mu_2(r+1)r^2}{\epsilon}$$

The other equations follow at once from $r = m_1' + m_2'$ and $\epsilon = m_1' m_2'$. The distance between the mode and mean is $a_1 - \mu_1' = a_1 - bm_1'/(m_1' + m_2')$, which can be easily reduced to the form given. A general value (regardless of type) for the distance was given in Chapter IV, § 5.

SECOND MAIN TYPE (TYPE IV)

$$y = y_0 \left(1 + \frac{x^2}{a^2}\right)^{-m} e^{-\nu \tan^{-1} x/a}$$

Origin is $\nu a/r$ after mean

The values to be calculated in order are

$$r = \frac{6(\beta_2 - \beta_1 - 1)}{2\beta_2 - 3\beta_1 - 6}$$

$$m = \tfrac{1}{2}(r + 2)$$

$$\nu = \frac{r(r-2)\sqrt{\beta_1}}{\sqrt{\{16(r-1) - \beta_1(r-2)^2\}}}$$

$$a = \sqrt{\frac{\mu_2}{16}} \sqrt{\{16(r-1) - \beta_1(r-2)^2\}}$$

$$y_0 = \frac{N}{a F(r, \nu)}$$

$$\text{Mode} = \text{mean} - \frac{1}{2}\frac{\mu_3(r-2)}{\mu_2(r+2)}$$

The curve is skew and has unlimited range in both directions. μ_3 and ν have opposite signs, i.e. when μ_3 is positive ν is negative.

A simple way to calculate the curve is to put it in the form

$$x = a\tan\theta, \quad y = y_0 \cos^{r+2}\theta\, e^{-\nu\theta}$$

Then θ is taken as $10°$, $20°$, $30°$, etc., and x and y found; this gives corresponding values of x and y, but the values of y will not be equidistant values of x. In calculating $e^{-\nu\theta}$ the value of θ must be taken in circular measure. If equidistant ordinates are required to be calculated accurately, little is gained by the double form, and if we had good tables of $\log(1+x^2)$ and $\tan^{-1} x$, the calculation of a particular ordinate would be a very simple matter. The calculation and meaning of $F(r, \nu)$ are dealt with in the proof. The log of this function is tabulated in *Tables for Statisticians*. When r is fairly large a close approximation to y_0, where $\tan\phi = \nu/r$, is given by

$$\frac{N}{a} \cdot \sqrt{\frac{r}{2\pi}} \cdot \frac{e^{\frac{\cos^2\phi}{3r} - \frac{1}{12r} - \phi\nu}}{(\cos\phi)^{r+1}}$$

We appear to reach the expression that looks shortest and simplest with the origin as shown on the previous page; it has generally been used and it is therefore given. This origin has, however, no physical meaning and there is much to be said for using the more complicated looking form with the origin at the mean, namely

$$y = y_0\left\{1 + \left(\frac{x}{a} - \frac{\nu}{r}\right)^2\right\}^{-m} e^{-\nu\tan^{-1}(x/a - \nu/r)}$$

see table facing p. 51.

The value of this expression when $x = 0$, i.e. the value of the ordinate at the mean, is

$$y_0\left\{1 + \frac{\nu^2}{r^2}\right\}^{-m} e^{-\nu\tan^{-1}(-\nu/r)} = \frac{N}{\sigma} \cdot \frac{1}{H(r, \nu)}$$

5-2

where $H(r, \nu)$ is a function related to $F(r, \nu)$. Its logarithm is also tabulated in *Tables for Statisticians*. The reader will appreciate at once that this curve needs considerable care; it is the most difficult of all the Pearson-type curves.

EXAMPLES

The numbers in the following nearly symmetrical distribution represent the exposed to risk of sickness by Sutton's Sickness Tables (males—all durations) when the number of weeks' sickness is represented by the normal curve of error.

Central age	No. exposed	Graduated by Type IV
5	10	6*
10	13	16
15	41	49
20	115	135
25	326	321
30	675	653
35	1,113	1,108
40	1,528	1,535
45	1,692	1,712
50	1,530	1,522
55	1,122	1,074
60	610	604
65	255	274
70	86	102
75	26	32
80	8	8
85	2	2
90	1	1
95	1	...
...	9,154	9,154

* This group has been taken as the area of the rest of the curve.

The following values were obtained:

$$\text{Mean} = 44 \cdot 5772339 \qquad \beta_1 = \cdot 0053656$$

$$\mu_2 = 4 \cdot 527608 \qquad \beta_2 = 3 \cdot 169897$$

$$\mu_3 = - \cdot 705687 \qquad \kappa = \cdot 0125$$

$$\mu_4 = 64 \cdot 98048$$

Type IV was used because, as there is a large number of cases, the standard error of κ will be small (see Chapter X)

$$r = 40{\cdot}12143$$
$$\nu = 4{\cdot}450398 \text{ (positive because } \mu_3 \text{ is negative)}$$
$$a = 13{\cdot}39152$$
$$m = 21{\cdot}06072$$
$$\text{Sk.} = -{\cdot}03313$$

When the 5-years unit with which we have been working is changed to one year, a becomes $66{\cdot}9576$, and $a^2 = 4483{\cdot}325$.

$$\text{The origin} = \text{mean} + \nu a/r$$
$$= 52{\cdot}504394$$

The mode, which is wanted if the curve is drawn, is at $44{\cdot}92989$.

As r is large the approximate form for y_0 was used, $\tan\phi = \dfrac{4{\cdot}450398}{40{\cdot}12143}$, or, $\log\tan\phi = \log\tan 6° 19' \dfrac{8925}{11537}$; hence $\log\cos\phi = \bar{1}{\cdot}9973446$, and from this y_0 is found to be $273{\cdot}3649$.

The value was checked by the tables in *Tables for Statisticians*.

The calculation of ordinates by the double process is as follows:

θ	x in years of age	$-4{\cdot}450398\,\theta\log e$	$42{\cdot}12143\log\cos\theta$	$\log y$	y
0^0	0	...	...	$2{\cdot}43675$	$273{\cdot}37$
1^0	$1{\cdot}1687$	$\bar{1}{\cdot}96637...$	$\bar{1}{\cdot}99721$	$2{\cdot}40033$	$251{\cdot}38$
2^0	$2{\cdot}3382$	$\bar{1}{\cdot}93253...$	$\bar{1}{\cdot}98885$	$2{\cdot}35813$	$228{\cdot}10$

The second column is formed directly from the tables of $\tan\theta$ by multiplying by a, and as x is required in years, $13{\cdot}39152 \times 5 = 66{\cdot}9576$ should be used for a. The fourth column is formed by multiplying $\log\cos\theta$ by $r + 2$, and the third continuously by addition. When θ is negative, the third column has to be subtracted from the fourth: i.e. it ceases to be negative and becomes positive. In each case the fifth is formed from the fourth + the third + $\log y_0$.

When drawing a curve of this type the position and height of the mode can be noted and then corresponding points inserted, e.g. $x = +1 \cdot 1687$ and $y = 251 \cdot 38$. Care must be taken to give the curve its maximum at the right point.

If the calculation is made directly, the following columns can be used:

x/a (1)	$1 + x^2/a^2$ (2)	$\log\,(1 + x^2/a^2)$ (3)	$\tan^{-1} x/a$ in degrees, etc. (4)	col. (4) in circular measure (5)	col. (5) $\times (-\nu\,\log_{10} e)$ (6)	$-m \times$ col. (3) (7)	$\log y_0$ $+(6)+(7)$ (8)	$y =$ antilog (8) (9)

Col. (2) can be formed by differences since $\Delta(1 + X^2) = 2X + 1$; $\tan^{-1} x/a$ has to be found by using a table of the tangents of angles inversely. A table helpful for obtaining col. (5) from col. (4) will be found in *Chambers' Mathematical Tables* or in *Tables for Statisticians*.

The troublesome work of inverse interpolation in degrees, minutes and seconds can be avoided by numbering the items in a table of $\tan \theta$ from 0 onwards. *Chambers' Tables*, for instance, give tangents for each minute in the following form:

1	0°	1°	2°, etc.
0	·0000000	(60)·0174551	(120)·0349208
1	·0002909	(61)·0177460	(121)·0352120
2	·0005818	(62)·0180370	(122)·0355033
3	·0008727	(63)·0183280	(123)·0357945
etc.			

If in the column headed 1° we insert 60, 61, etc., and in the column headed 2° we insert 120, 121, etc., as indicated by the figures in brackets, we can make the inverse interpolation in minutes. Then, as one minute in circular measure is ·0002908882, we can obtain the figure we require by multiplying by the conversion factor. In practice however it would be combined into one multiplier with $(-\nu \log_{10} e)$ and col. (6) would be found directly from col. (4) by multiplying, in our example, by ·003519003. The labour of inserting the minutes in a printed table is small, as all we need to do is to write the number of minutes under the number of degrees at the head of each column and add thereto at sight the marginal minutes when the interpolations are being made.

Tables of $\tan^{-1}\theta$ and $\log(1+x^2)$ are given in *Tracts for Computers*, No. XXIII and these tables simplify the calculations.

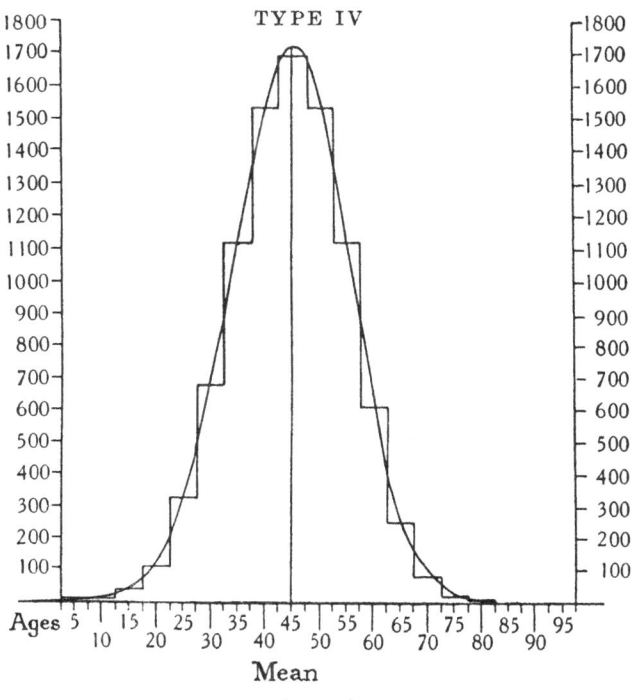

TYPE IV

Ages

Mean

In $\qquad y = y_0 \left\{1 + \dfrac{x^2}{a^2}\right\}^{-m} e^{-\nu \tan^{-1} x/a}$ put $\tan\theta = \dfrac{x}{a}$

$\therefore \qquad \theta = \tan^{-1}\dfrac{x}{a}$ and

$\left\{1 + \left(\dfrac{x}{a}\right)^2\right\}^{-m} = \{1 + \tan^2\theta\}^{-m} = (\sec^2\theta)^{-m} = \cos^{2m}\theta$

$\therefore \qquad y = y_0 \cos^{2m}\theta\, e^{-\nu\theta}$

Now $\qquad N = \displaystyle\int_{-\infty}^{+\infty} y_0 \left\{1 + \dfrac{x^2}{a^2}\right\}^{-m} e^{-\nu\tan^{-1}x/a}\, dx$

$\qquad\qquad = \displaystyle\int_{-\frac{1}{2}\pi}^{\frac{1}{2}\pi} y_0 \cos^{2m}\theta\, e^{-\nu\theta} \dfrac{a}{\cos^2\theta}\, d\theta$, by substituting

$\tan\theta = \dfrac{x}{a}$ so that $\dfrac{dx}{d\theta} = a\sec^2\theta = \dfrac{a}{\cos^2\theta}$

$\qquad\qquad = y_0 a \displaystyle\int_{-\frac{1}{2}\pi}^{\frac{1}{2}\pi} \cos^r\theta\, e^{-\nu\theta}\, d\theta$ where $r = 2m - 2$

$\qquad\qquad = y_0 a e^{-\frac{1}{2}\nu\pi} \displaystyle\int_0^\pi \sin^r\phi\, e^{\nu\phi}\, d\phi,$

substituting $\sin\phi$ for $\cos\theta$ so that $\frac{1}{2}\pi = \theta + \phi$ and changing limits, $= y_0 a F(r, \nu)$, say.

The nth moment about the origin is

$\mu_n' = \dfrac{1}{N} \displaystyle\int_{-\infty}^{\infty} y x^n\, dx$

$\quad = \dfrac{1}{N} \displaystyle\int_{-\infty}^{\infty} y_0 x^n \left\{1 + \dfrac{x^2}{a^2}\right\}^{-m} e^{-\nu\tan^{-1}x/a}\, dx$

$\quad = \dfrac{1}{N} \displaystyle\int_{-\frac{1}{2}\pi}^{\frac{1}{2}\pi} y_0 a^{n+1} \cos^{2m-2}\theta \tan^n\theta\, e^{-\nu\theta}\, d\theta$, by substituting as above,

$\quad = \dfrac{y_0 a^{n+1}}{N} \displaystyle\int_{-\frac{1}{2}\pi}^{\frac{1}{2}\pi} \cos^{r-n}\theta \sin^n\theta\, e^{-\nu\theta}\, d\theta$

$\quad = \dfrac{y_0 a^{n+1}}{N} \left[-\dfrac{\cos^{r-n+1}\theta \sin^{n-1}\theta\, e^{-\nu\theta}}{r-n+1} \right.$

$\qquad\qquad \left. + \displaystyle\int \left\{\dfrac{\cos^{r-n+1}\theta}{r-n+1} [\sin^{n-2}\theta\cos\theta\, e^{-\nu\theta}(n-1) - \nu e^{-\nu\theta}\sin^{n-1}\theta] \right\} d\theta \right]_{-\frac{1}{2}\pi}^{\frac{1}{2}\pi}$

by integrating by parts and treating $\sin^{n-1}\theta e^{-\nu\theta}$ as one part and $\cos^{r-n}\theta\sin\theta$ as the other, and remembering that

$$\int \cos^{r-n}\theta\sin\theta\, d\theta = -\frac{\cos^{r-n+1}\theta}{r-n+1}.$$

Now, since $\cos^{r-n+1}\theta\sin^{n-1}\theta e^{-\nu\theta} = 0$ when θ becomes $\frac{1}{2}\pi$ or $-\frac{1}{2}\pi$, we have

$$\mu_n' = \frac{y_0 a^{n+1}}{N(r-n+1)}\int_{-\frac{1}{2}\pi}^{\frac{1}{2}\pi}\{(n-1)\cos^{r-n+2}\theta\sin^{n-2}\theta e^{-\nu\theta}$$

$$- \nu\cos^{r-n+1}\theta\sin^{n-1}\theta e^{-\nu\theta}\}\, d\theta$$

$$= \frac{a}{r-n+1}\{(n-1)a\mu_{n-2}' - \nu\mu_{n-1}'\}$$

Further,

$$\mu_1' = \frac{y_0 a^2}{N}\int_{-\frac{1}{2}\pi}^{\frac{1}{2}\pi}\cos^r\theta\tan\theta e^{-\nu\theta}\, d\theta$$

$$= \frac{y_0 a^2}{Nr}\left\{-\int_{-\frac{1}{2}\pi}^{\frac{1}{2}\pi}\nu\cos^r\theta e^{-\nu\theta}\, d\theta\right\}$$

by putting $n = 1$ in the above equation for μ_n'

$$= -\frac{a\nu}{r}, \text{ because } N = y_0 a\int_{-\frac{1}{2}\pi}^{\frac{1}{2}\pi}\cos^r\theta e^{-\nu\theta}\, d\theta$$

Using the last result with the formula for the nth in terms of the two previous moments, and remembering that μ_0' is unity,

$$\mu_2' = \frac{a^2}{r(r-1)}(r+\nu^2)$$

$$\mu_3' = -\frac{a^3\nu}{r(r-1)(r-2)}(3r-2+\nu^2)$$

$$\mu_4' = \frac{a^4}{r(r-1)(r-2)(r-3)}\{3r(r-2)+\nu^2(6r-8)+\nu^4\}$$

Referring these moments to the centroid vertical, we have, by putting $d = \mu_1' = -\dfrac{a\nu}{r}$ in the formulae on p. 57,

(73)

$$\mu_2 = \frac{a^2}{r^2(r-1)}\,(r^2 + \nu^2)$$

$$\mu_3 = -\frac{4a^3\nu(r^2 + \nu^2)}{r^3(r-1)(r-2)}$$

$$\mu_4 = \frac{3a^4(r^2 + \nu^2)\{(r+6)(r^2 + \nu^2) - 8r^2\}}{r^4(r-1)(r-2)(r-3)}$$

If now, we put z for $r^2 + \nu^2$, and write as before,

$$\beta_1 = \frac{\mu_3^2}{\mu_2^3} \quad \text{and} \quad \beta_2 = \frac{\mu_4}{\mu_2^2}$$

we have

$$-\frac{\beta_1(r-2)^2}{2(r-1)} = 8\frac{r^2}{z} - 8$$

and

$$\frac{\beta_2(r-2)(r-3)}{3(r-1)} = r + 6 - \frac{8r^2}{z}$$

Adding and dividing out by $r - 2$, we have

$$r = \frac{6(\beta_2 - \beta_1 - 1)}{2\beta_2 - 3\beta_1 - 6}$$

and

$$z = \frac{r^2}{1 - \dfrac{\beta_1(r-2)^2}{16(r-1)}}$$

Finally, since $\nu^2 = z - r^2$, the other formulae on p. 66 follow at once.

Since the tangent at the top of the maximum ordinate is parallel to the axis of x, the position of the mode is such that dy/dx is zero at that point, i.e.

$$y_0\left\{1 + \frac{x^2}{a^2}\right\}^{-(m+1)} e^{-\nu\tan^{-1}x/a}\left[-\frac{2mx}{a^2} - \frac{\nu}{a}\right]$$

is zero. There are three cases, $x = -\infty$, $x = +\infty$, and a value of x such that $\dfrac{2mx}{a^2} + \dfrac{\nu}{a}$ is zero, or $x = -\dfrac{\nu a}{2m}$. The distance of the mean from the origin is μ_1' or $-\dfrac{\nu a}{r}$, and, therefore, the distance

(74)

between the mean and mode is $-\dfrac{2\nu a}{r(r+2)}$, which reduces to the expression given on p. 66, when the values for ν and a, on the same page, are inserted.

It will be useful to give another example of the calculation of y_0 for curves of this type, and we may take a curve where $r = 29\cdot590$, $\nu = 19\cdot886$, $a = 13\cdot650$, $N = 2162$. Hence $\tan\phi = \cdot67205$, $\phi = 33° 54' \tfrac{8}{42}$, $\cos\phi = \cdot82998$, $\log\cos\phi = \overline{1}\cdot91907$, and ϕ in circular measure is $\cdot59172$.

$$\log N = 3\cdot33486$$
$$\operatorname{colog} a = \overline{2}\cdot86486$$
$$\tfrac{1}{2}\log r = \cdot73557$$
$$\log\frac{1}{\sqrt{2\pi}} = \overline{1}\cdot60091$$

$$\frac{\cos^2\phi}{3r} = \cdot00776$$

$$-\frac{1}{12r} = -\ \cdot00282$$

$$-\phi\nu = -11\cdot76700$$

$$\overline{-11\cdot762} \ \times\log_{10} e = \overline{6}\cdot89183$$
$$\operatorname{colog}(\cos\phi)^{r+1} = \overline{2}\cdot47564$$
$$\overline{\overline{1}\cdot90367}$$
$$y_0 = \overline{\cdot80107}$$

The form just considered is sufficiently accurate for all practical purposes provided ν is not very small. If $\nu < 2$ the tables in *Tables for Statisticians* must be used.

THIRD MAIN TYPE (TYPE VI)

$$y = y_0 \, (x-a)^{q_2} x^{-q_1}$$

Origin at a before start of curve

The values to be calculated in order are

$$r = \frac{6(\beta_2 - \beta_1 - 1)}{6 + 3\beta_1 - 2\beta_2}$$

$$a = \frac{1}{2} \sqrt{\mu_2} \sqrt{\{\beta_1(r+2)^2 + 16(r+1)\}}$$

q_2 and $-q_1$ are given by

$$\frac{r-2}{2} \pm \frac{r(r+2)}{2} \sqrt{\frac{\beta_1}{\beta_1(r+2)^2 + 16(r+1)}}$$

$$y_0 = \frac{N a^{q_1 - q_2 - 1} \, \Gamma(q_1)}{\Gamma(q_1 - q_2 - 1) \, \Gamma(q_2 + 1)}$$

$$\text{Origin} = \text{Mean} - \frac{a(q_1 - 1)}{q_1 - q_2 - 2}$$

$$\text{Mode} = \text{Mean} - \frac{1}{2} \cdot \frac{\mu_3}{\mu_2} \cdot \frac{r+2}{r-2}$$

If expressing curve with origin at mean (see Table VI, facing p. 51):

$$A_1 = \frac{a(q_1 - 1)}{(q_1 - 1) -- (q_2 + 1)}, \quad A_2 = \frac{a(q_2 + 1)}{(q_1 - 1) - (q_2 + 1)}$$

$$y_e = \frac{N(q_2 + 1)^{q_2} (q_1 - q_2 - 2)^{q_1 - q_2} \, \Gamma(q_1)}{a(q_1 - 1)^{q_1} \, \Gamma(q_1 - q_2 - 1) \, \Gamma(q_2 + 1)}$$

NOTES

The range is from a to ∞ and the method is like that of Type I. If μ_3 is negative, then a is negative and the range is from $-\infty$ to $-a$.

r is always negative and q_1 is greater than q_2. If q_2 is negative, the curve is J-shaped.

The value of y_0 does not correspond to any frequency, as it relates to a point before the curve starts.

The reader will probably find it easier to work with the origin at the mean, and in the numerical example both forms are shown.

EXAMPLE

The number of entrants, limited payment policies, 1863–93 experience was summed in groups of ten years of age and divided by 100, and the following series was obtained:

No. of entrants ÷100	Graduated by Type VI curve
1	1
56	50
167	168
98	100
34	36
9	10
2	2
1	·5
368	368

The moments, etc. were

Mean at ·402174 after the centre of 167 group

$\mu_2 =$ ·928835 $1 - q_1 = -41\cdot03080$

$\mu_3 =$ ·893096 $1 + q_2 = 7\cdot60950$

$\mu_4 =$ 4·088800 $q_1 = 42\cdot03080$

$\beta_1 =$ ·9953605 $q_2 = 6\cdot60950$

$\beta_2 =$ 4·739349 $a = 10\cdot37947$

$\kappa =$ 1·895 $\log y_0 = 46\cdot1821$

$r = -33\cdot42129$

(77)

The origin is 12·74270 before the mean or 12·34053 before the centre of the 167 group, and the curve starts at 12·34053 − 10·37949 = 1·96106 before the centre of the largest group. This makes the start of the curve at about age 10, which is reasonable.

If we use the origin at the mean, we have

$$A_1 = 12\cdot74270, \quad A_2 = 2\cdot36324, \quad y_e = 147\cdot4$$

and the range is from − 2·36324 to ∞.

The curve was calculated as follows:

x (1)	$\log x$ (2)	$\log(x-a)$ (3)	$-q_1 \log x$ (4)	$q_2 \log(x-a)$ (5)	$\log y$ (6)	y (7)

There is no difficulty in writing down the values for cols. (2) and (3) without using col. (1), as only the whole numbers in x and $x-a$ change, the decimal remaining constant so long as equidistant ordinates are required. Cols. (4) and (5) are obtained directly, and col. (6) by adding cols. (4) and (5) to $\log y_0$. Cols. (2) and (3) can be formed continuously with the aid of Gauss-logarithms.

The mode which is useful for drawing the curve is ·02429 before the centre of the largest group.

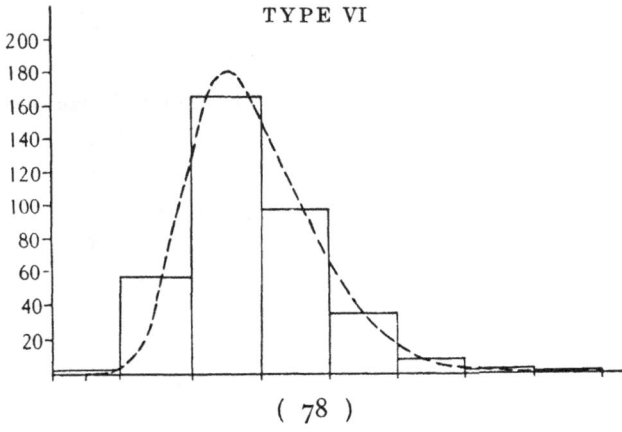

TYPE VI

With the origin at the mean the form of the columns is similar to that already shown for Type I.

PROOF

$$N = \int_a^\infty y_0 (x-a)^{q_2} x^{-q_1} \, dx$$

$$= \int_a^\infty y_0 a^{q_2-q_1} \left(\frac{x}{a} - 1\right)^{q_2} \left(\frac{x}{a}\right)^{-q_1} dx$$

$$= \int_1^0 y_0 a^{q_2-q_1} \left(\frac{1}{z} - 1\right)^{q_2} z^{q_1} (-a z^{-2}) \, dz$$

by substituting $1/z$ for x/a

$$= \int_0^1 y_0 a^{q_2-q_1+1} (1-z)^{q_2} z^{q_1-q_2-2} \, dz$$

$$\therefore \quad y_0 = \frac{N}{a^{q_2-q_1+1} \, B(q_2+1, \, q_1-q_2-1)}$$

$$= \frac{N\Gamma(q_1)\, a^{q_1-q_2-1}}{\Gamma(q_2+1)\,\Gamma(q_1-q_2-1)}$$

The nth moment about the origin is

$$\mu_n' = \frac{1}{N} \int_a^\infty y_0 x^n (x-a)^{q_2} x^{-q_1} dx$$

$$= \frac{y_0}{N a^{q_1-q_2-n-1}} \frac{\Gamma(q_1-q_2-n-1)\,\Gamma(q_2+1)}{\Gamma(q_1-n)}$$

by the same substitution as that used above.

From this last result we obtain, by inserting the value of y_0, and remembering the relationship between $\Gamma(q_1)$ and $\Gamma(q_1-1)$, etc.,

$$\mu_1' = \frac{a(q_1-1)}{q_1-q_2-2}$$

$$\mu_2' = \frac{a^2(q_1-1)\,(q_1-2)}{(q_1-q_2-2)\,(q_1-q_2-3)}$$

etc.

It will be noticed that these equations are the same as those obtained for Type I if $m_1 = -q_1$, $m_2 = q_2$ and $b = a$. Thus, we can use the whole of the Type I solution, provided we bear in mind that the range is from $x = a$ to $x = \infty$.

(79)

TRANSITION TYPE

"NORMAL CURVE OF ERROR"

$$y = y_0 e^{-x^2/c}$$

$$c = 2\sigma^2$$

$$y_0 = \frac{N}{\sqrt{(2\pi\mu_2)}}$$

This curve has been known by various names, such as the Probability Curve and the Gaussian Curve. It was discussed before Gauss by de Moivre and Laplace. It is the limit of $(p+q)^n$ where $p+q = 1$, when n approaches infinity and if neither p nor q is very small. It gives a close representation of $(\frac{1}{2} + \frac{1}{2})^n$ even when n is not large.

EXAMPLES

The following table gives, in col. (2), the sums assured and bonuses, and in col. (4) the reserves resulting from grouping a number of Endowment Assurances according to their office years of birth:

Central age for groups of 5 years of birth	Sums assured and bonuses/1,000		Reserves/1,000	
	Ungraduated	Graduated	Ungraduated	Graduated
(1)	(2)	(3)	(4)	(5)
17	11	16	·6	·6
22	48	45	2·8	2·8
27	124	115	11·5	10·9
32	213	213	27·7	30·1
37	281	287	59·1	58·4
42	295	282	84·7	79·9
47	185	202	74·1	77·0
52	104	105	50·5	52·2
57	40	40	23·2	25·0
62	15	11	12·2	8·4
67	3	3	1·3	2·4
Total ...	1,319	1,319	347·7	347·7

The following table shows the moments and constants:

Constant	Sum assured and bonus	Reserves
Mean age	39·202426	43·967213
μ_2	3·066840	2·769635
μ_3	·650127	·029805
μ_4	27·02516	22·4\u0020663
β_1	·014653	·0000418
β_2	2·873346	2·920997
κ	$-$·04	$-$·0002
$\sigma(=\sqrt{\mu_2})$	1·751237	1·664222
σ^{-1}	·5710248	·6008813
y_0	300·4760	83·34959

The criteria for the normal curve are $\kappa = 0$, $\beta_1 = 0$, and $\beta_2 = 3$. The values given above do not differ very greatly from these, but a comparison of the graduated and ungraduated figures shows that the reserve curve agrees better than the sum assured curve; partly because the value of β_2 is closer to 3, and β_1 has a larger value in the case of the sum assured.

For the calculation of y_0 the value of

$$\operatorname{colog} \sqrt{2\pi} = \bar{1}{\cdot}6009100657$$

is required.

In finding the areas for the comparison between the graduated and ungraduated figures it is unnecessary to calculate the ordinates, as one of the calculated tables of the probability integral can be used. The table by W. F. Sheppard included in *Tables for Statisticians* is very convenient, and the columns in the table below show how it was used to calculate

Age x	Distance from origin in calculation units, i.e. 5 years of age	Previous column $\times \sigma^{-1}$	Values of $\frac{1}{2}(1+\alpha)$ from Sheppard's tables using differences (area from origin to x)	Difference of previous column = area for age group x to $x+5$	Area multiplied by 347·7 (total) frequency)
14·5	5·893443	3·541258	...	·00164*	·6
19·5	4·893443	2·940377	·99836	·00802	2·8
24·5	3·893443	2·339496	·99034	·03139	10·9
29·5	2·893443	1·738615	·95895	·08657	30·1
34·5	1·893443	1·137734	·87238	·16806	58·4
39·5	·893443	·536853	·70432	·22985†	79·9
44·5	·106557	·064028	·52553	·22141	77·0
49·5	1·106557	·664909	·74694	·15018	52·2
54·5	2·106557	1·265790	·89712	·07190	25·0
59·5	3·106557	1·866671	·96902	·02418	8·4
64·5	4·106557	2·467552	·99320	·00572	2·0
69·5	5·106557	3·068443	·99892	·00108*	·4

* Remainders of areas beyond 19·5 and 69·5.

† $(\cdot70432 - \cdot50000) + (\cdot52553 - \cdot50000)$ because we pass across the origin, and a piece of the group is on each side of it.

the areas in one of the cases (the reserves). Sheppard's tables give the areas and ordinates of the normal curve in terms of the standard deviation; that is, he assumes the standard deviation to be unity, and his tables must be entered by using intervals of σ^{-1}. A short abstract from Sheppard's table is given on p. 265. Most other published tables are based on the standard deviation multiplied by $\sqrt{2}$ and the distinction must be borne in mind if other tables are used.

The second column can be left out when the method has been grasped. The ages in the first column were taken consistently with the assumptions that 17, 22, etc. were the central ages of the groups.

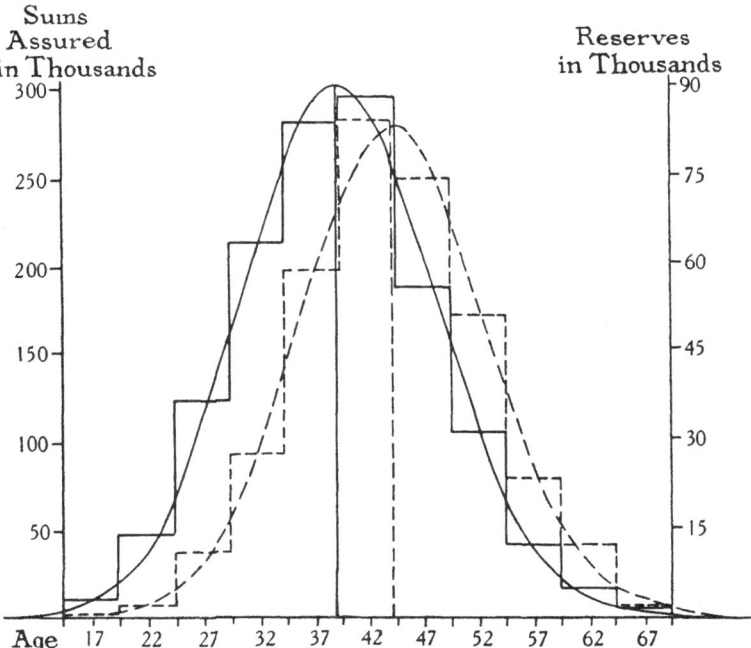

"NORMAL CURVE OF ERROR"

If ordinates are required, the z column in Sheppard's tables must be used. It was with its help that the curves in the figure were drawn. The statistics and curve for the reserves are shown by the dotted lines.

(83)

An average reserve for any group can be obtained by means of the graduated figures, and it could be used to test the reserves obtained at any future valuation. This is by no means the only rough check that can be applied, but it is interesting because it shows a use to which frequency-curves might be put in practical office routine.

PROOF

To show that

$$\int_0^\infty e^{-x^2}dx = \frac{\sqrt{\pi}}{2}$$

let

$$\int_0^\infty e^{-x^2}dx = \kappa$$

then, substituting ax for x, we have

$$\int_0^\infty e^{-a^2x^2}a\,dx = \kappa$$

$\therefore$

$$\int_0^\infty e^{-a^2(1+x^2)}a\,dx = \kappa e^{-a^2}$$

Hence

$$\int_0^\infty \int_0^\infty e^{-a^2(1+x^2)}a\,da\,dx = \kappa \int_0^\infty e^{-a^2}da = \kappa^2$$

But

$$\int_0^\infty e^{-a^2(1+x^2)}a\,da = \frac{1}{2}\cdot\frac{1}{1+x^2}$$

$\therefore$

$$\frac{1}{2}\int_0^\infty \frac{dx}{1+x^2} = \kappa^2$$

and

$$\kappa^2 = \frac{\pi}{4} \text{ or } \kappa = \frac{\sqrt{\pi}}{2}$$

Hence

$$\int_{-\infty}^\infty e^{-x^2}dx = \sqrt{\pi}$$

(84)

The other constant is obtained as follows:

$$\int_{-\infty}^{\infty} y_0 e^{-x^2/c}\, dx = y_0 \left[x e^{-x^2/c} + \int \frac{2x}{c} e^{-x^2/c}\, x\, dx \right]_{-\infty}^{\infty} \text{ by parts}$$

$$= \frac{2y_0}{c} \int_{-\infty}^{\infty} x^2 e^{-x^2/c}\, dx$$

or $$N = \frac{2N}{c} \mu_2$$

$$\therefore \qquad c = 2\mu_2$$

TRANSITION TYPE (TYPE II)

$$y = y_0 \left(1 - \frac{x^2}{a^2}\right)^m$$

Origin at mode (= mean)

$$m = \frac{5\beta_2 - 9}{2(3 - \beta_2)}$$

$$a^2 = \frac{2\mu_2\beta_2}{3 - \beta_2}$$

$$y_0 = \frac{N \times \Gamma(2m + 2)}{a \times 2^{2m+1}\{\Gamma(m + 1)\}^2}$$

$$= \frac{N}{a\sqrt{\pi}} \cdot \frac{\Gamma(m + 1\frac{1}{2})}{\Gamma(m + 1)}$$

Put $\beta_1 = 0$ in Type I, for the curve is symmetrical, and therefore $\mu_3 = 0$. For the same reason it is clear that $m_1 = m_2$. Approximations to Γ may be used if m is large.

If m is positive, the curve starts at zero, rises to a maximum and falls again to zero; but if m is negative, it starts at infinity, falls, and then rises to infinity again.

EXAMPLE

In the discussion that followed the reading of G. J. Lidstone's paper on Endowment Assurances, G. F. Hardy said that "the errors in the successive groups formed a curve very similar to the normal curve of error" (*J. Inst. Actu.* XXXIV, 87), and the series in question is a rather interesting example of a symmetrical distribution.

Unexpired term in years	Error involved in using "mean age" method
0–4	11
5–9	116
10–14	274
15–19	451
20–24	432
25–29	267
30–34	116
35, etc.	16
...	1,683

Moments were calculated about the centre of the 15–19 group, and ·4985146, 2·161022, 3·104576, and 12·60666 were found for the first four moments; transferring to the mean ($17·5 + 2·492573 = 19·992573$), and using Sheppard's adjustments, the following values result:

$$\mu_2 = 1·829172 \qquad \beta_1 = \quad ·0023706$$
$$\mu_3 = \quad ·120452 \qquad \beta_2 = \quad 2·548313$$
$$\mu_4 = 8·52636 \qquad \kappa = -\ ·007492$$

which shows that Type II can be used.

The equations for the type give

$$m = 4\cdot141766, \quad a = 4\cdot543079, \quad y_0 = 462\cdot57$$

The mean and mode coincide, because the curve is symmetrical.

For calculating a series of values, the following arrangement is convenient:

$\dfrac{x}{a}$	$\log\left(1 + \dfrac{x}{a}\right)$	$\log\left(1 - \dfrac{x}{a}\right)$	$(2) + (3)$	$\log y_x$ $= m \times (4)$ $+ \log y_0$
(1)	(2)	(3)	(4)	(5)

It is easier to work in this way than by calculating values of $1 - x^2/a^2$. In the particular example, ordinates were calculated at the beginning, middle, and end of each group, and Simpson's quadrature formula was used for finding the areas, viz.

$$\int_0^1 y\,dx = \tfrac{1}{6}\{y_0 + 4y_{\frac{1}{2}} + y_1\}.$$

Group	Ungraduated figures	Areas Type II	Mid-ordinates, Type II	Areas, "Normal curve"
0–4	11	14	11	22
5–9	116	109	104	95
10–14	274	286	287	270
15–19	451	433	440	455
20–24	432	433	440	455
25–29	267	285	287	269
30–34	116	109	104	95
35, etc.	16	14	11	22
...	1,683	1,683	...	1,683

A comparison of the mid-ordinates with the areas gives an idea of the error involved in using the former for the latter; the differences are largest at the "tails" and near the mode.

The curve starts at $19 \cdot 992573 - 22 \cdot 71540 = -2 \cdot 72283$, and ends at $42 \cdot 70797$.

The final column of the table gives a graduation by the "normal curve".

TYPE II

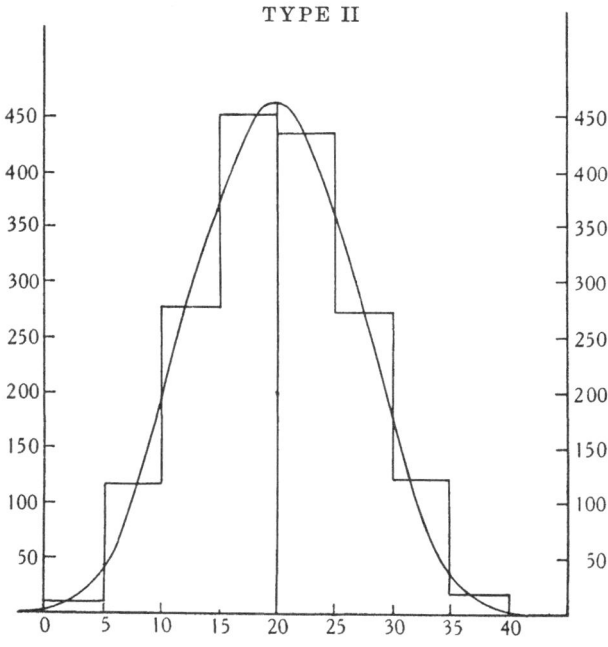

TRANSITION TYPE (TYPE VII)

$$y = y_0 \left(1 + \frac{x^2}{a^2}\right)^{-m}$$

Origin at mode (= mean)

$$m = \frac{5\beta_2 - 9}{2(\beta_2 - 3)}$$

$$a^2 = \frac{2\mu_2\beta_2}{\beta_2 - 3}$$

$$y_0 = \frac{N}{a\sqrt{\pi}} \cdot \frac{\Gamma(m)}{\Gamma(m - \frac{1}{2})}$$

The curve may be taken as a special case of Type IV when $\nu = 0$, or it can be evolved from Type II by making both m and a^2 negative in that type. This happens when $\beta_2 > 3$. The curve is symmetrical and of unlimited range in both directions.

$$N = \int_{-\infty}^{+\infty} y_0 \left(1 + \frac{x^2}{a^2}\right)^{-m} dx$$

$$= 2\int_0^\infty y_0 \left(1 + \frac{x^2}{a^2}\right)^{-m} dx$$

then putting $1 + \dfrac{x^2}{a^2} = z^{-1}$, the reader will be able to show that

$$N = \int_0^1 y_0 a (1-z)^{-\frac{1}{2}} z^{m-1\frac{1}{2}} dz$$

$$= a y_0 B(m - \tfrac{1}{2}, \tfrac{1}{2})$$

or y_0 has the value shown on the preceding page, because $\Gamma(\tfrac{1}{2}) = \sqrt{\pi}$.

EXAMPLE

The following table gives the areas when $\beta_2 = 5$ and $\mu_2 = 1$ and shows a graduation by the "normal curve". The example, together with that of Type II, will act as a reminder that the "normal curve" does not give entirely satisfactory results even with symmetrical distributions.

Type VII $m=4$, $a^2=5$	Normal curve $\sigma=1$
1	...
1	...
2	...
4	...
7	1
16	5
38	24
93	93
225	278
527	656
1,106	1,210
1,858	1,746
2,244	1,974
1,858	1,746
etc.	etc.

TRANSITION TYPE (TYPE III)

$$y = y_0\left(1 + \frac{x}{a}\right)^{\gamma a} e^{-\gamma x}$$

Origin at mode

$$\gamma = \frac{2\mu_2}{\mu_3}$$

$$p = \gamma a = \frac{4}{\beta_1} - 1$$

$$a = \frac{2\mu_2^2}{\mu_3} - \frac{\mu_3}{2\mu_2}$$

$$y_0 = \frac{N}{a} \cdot \frac{p^{p+1}}{e^p \, \Gamma(p+1)}$$

$$\text{Mode} = \text{Mean} - \frac{\mu_3}{2\mu_2}$$

If expressing curve with origin at mean (see Table VI, facing p. 51):

$$y_e = N \cdot \gamma \cdot \frac{(p+1)^p}{e^{p+1} \, \Gamma(p+1)}$$

The curve is usually bell-shaped, but becomes J-shaped when $p < 0$, that is, when $\beta_1 > 4$. The range is limited in one direction only. The criterion is that $2\beta_2 = 6 + 3\beta_1$. Theoretically this gives $\kappa = \infty$ but the curve may be used in many cases where κ is not very large, provided $2\beta_2$ approximates to $6 + 3\beta_1$. When μ_3 is positive γ and a are positive, so that the range is limited at a distance of a before the mode; when μ_3 is negative γ and a are negative, so that the range is limited at a distance a after the mode. If, however, $\beta_1 > 4$, then a and γ have different signs.

EXAMPLE

The following statistics are taken from a paper in the *Trans. Actu. Soc. Edinb.* IV, 44, and give the numbers of wives tabulated for the ages of mothers, and according to years since marriage. The mothers' ages for the particular series are 30 to 34.

Year after marriage	Number of wives	Graduated by Type III curve
1	44	59
2	135	111
3	45	45
4	12	20
5	8	9
6	3	4
7	1	2
8	3	1
Total ...	251	251

The mean is ·3346612 after the middle of the second group, and the moments about the centroid vertical are 1·441787, 3·606622 and 18·93221; so that $\kappa = -8·44$.

As this value was large, Type III was used, and

$$\gamma = \cdot7995221 \qquad a = -\cdot098007$$
$$p = -\cdot0783584 \qquad y_0 = 214\cdot8$$

This example is given because it can be used to show a difficulty rather clearly. At first sight, a curve starting at zero, rising to a maximum, and then falling, might be expected. Instead, we find the curve starting at duration ·68192;* so that

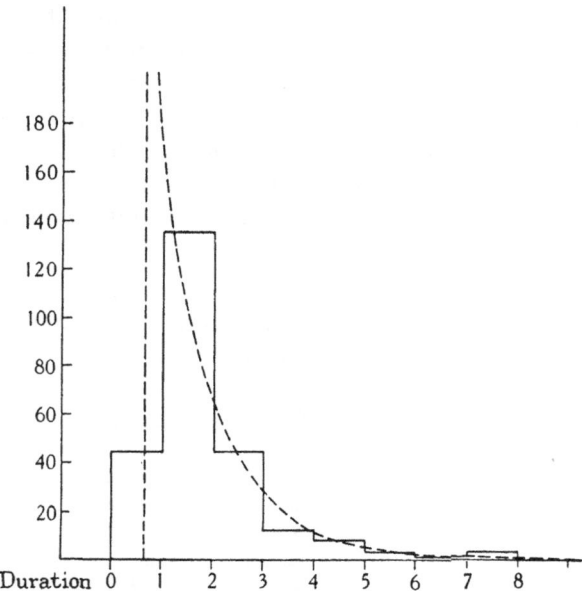

the first group is made up of a strip on a base ·31808 in length, and has a smaller value than the next group, though any ordinate read off within the first group would be larger than any ordinate in the second group. No adjustment was made to the rough moments.

* The mode in ordinary cases of Type III is given by mean $-\frac{\mu_3}{2\mu_2}$. In this case, $\frac{\mu_3}{2\mu_2} = 1\cdot25075$; so the mode would be at ·58391, and the curve would start at $\{\text{"mode"} - a\} = \cdot58391 + \cdot09801 = \cdot68192$.

PROOF

In the equation for the type, viz. $y = y_0 \left(1 + \dfrac{x}{a}\right)^{\gamma a} e^{-\gamma x}$, put $\gamma a = p$, and substitute z for $\gamma(a + x)$; then, if N be the total frequency,

$$N = \int_{-a}^{\infty} y_0 \left(1 + \frac{x}{a}\right)^p e^{-\gamma x}\, dx$$

$$= \int_{0}^{\infty} y_0 z^p a^{-p} e^{-z+p} \gamma^{-(p+1)}\, dz \quad \text{for } \frac{dz}{dx} = \gamma$$

$$= y_0 \frac{e^p}{\gamma p^p} \int_{0}^{\infty} z^p e^{-z}\, dz$$

$$= y_0 \frac{a e^p}{p^{p+1}} \Gamma(p+1)$$

This gives $y_0 = \dfrac{N p^{p+1}}{a e^p \Gamma(p+1)}$

The nth moment about the start of the curve is

$$\frac{1}{N} \int_{-a}^{\infty} y_0 \left(1 + \frac{x}{a}\right)^p e^{-\gamma x} (x+a)^n\, dx = \frac{y_0 e^p}{N p^p \gamma^{n+1}} \int_{0}^{\infty} z^{p+n} e^{-z}\, dz$$

$$= \frac{\Gamma(p+n+1)}{\gamma^n \Gamma(p+1)}$$

by using the value of y_0 found above.

Since $\Gamma(p) = (p-1)\,\Gamma(p-1)$, the first moment is $\dfrac{p+1}{\gamma}$, the second $\dfrac{(p+1)(p+2)}{\gamma^2}$, and the third $\dfrac{(p+1)(p+2)(p+3)}{\gamma^3}$ In order to apply these formulae to statistical work, it is necessary to have moments about the centroid vertical, the position of which (the mean) can be found; and as, by definition, the first moment about it is zero, we get

$$\mu_2 = \frac{p+1}{\gamma^2} \quad \text{and} \quad \mu_3 = \frac{2(p+1)}{\gamma^3}$$

These results give γ and p as $\dfrac{2\mu_2}{\mu_3}$ and $\dfrac{4\mu_2^3}{\mu_3^2} - 1$ respectively.

(95)

TRANSITION CURVE (TYPE V)

$$y = y_0 x^{-p} e^{-\gamma/x}$$

Origin at start of curve

$$p = 4 + \frac{8 + 4\sqrt{\{4 + \beta_1\}}}{\beta_1}$$

$$\gamma = (p - 2)\sqrt{\{\mu_2(p - 3)\}}$$

$$y_0 = \frac{N\gamma^{p-1}}{\Gamma(p - 1)}$$

$$\text{Origin} = \text{Mean} - \frac{\gamma}{p - 2}$$

$$\text{Mode} = \text{Mean} - \frac{2\gamma}{p(p - 2)}$$

The sign of γ is the same as that of μ_3.

If expressing curve with origin at mean (see Table VI, facing p. 51):

$$A = \gamma/(p - 2)$$

$$y_e = \frac{N(p - 2)^p}{\gamma e^{p-2}\Gamma(p - 1)}$$

The following series of deaths is taken from G. King's paper "On the rate of mortality amongst female nominees, etc." (*J. Inst. Actu.* xxxiii, 262–8):

Ages	Deaths	Graduated by Type V
30–34	1	1
35–39	5	3
40–44	8	6
45–49	12	14
50–54	28	32
55–59	82	68
60–64	128	137
65–69	253	247
70–74	342	381
75–79	525	480
80–84	438	441
85–89	265	261
90–94	53	80
95–99	18	10
100, etc.	4	1
...	2,162	2,162

The mean is at age 75·9782605, and the moments (adjusted), etc. are

$$\mu_2 = \quad 3\cdot573346 \qquad\qquad \beta_1 = \quad \cdot4950399$$

$$\mu_3 = -4\cdot752613 \qquad\qquad \beta_2 = 3\cdot996134$$

$$\mu_4 = \quad 51\cdot02583 \qquad\qquad \kappa = \quad \cdot85$$

Strictly speaking, Type IV should be used, but the value is not very far from unity, and the following Type V constants were found:

$$p = \quad 37\cdot29145$$

$$\gamma = -390\cdot6609 \text{ (negative, because } \mu_3 \text{ is)}$$

$$\log y_0 = \quad 56\cdot930518$$

The approximation to the value of $\log \Gamma(p-1)$ was used. The origin is at age 131·32606, and the mode at 78·9467.

The columns used for calculating the ordinates were:

x (1)	$\log x$ (2)	$-p \log x$ (3)	$\dfrac{1}{x}(-\gamma \log_{10} e)$ (4)	$\begin{aligned}&\log y\\ =&\log y_0 + (3) + (4)\end{aligned}$ (5)	$y = $ antilog (5) (6)

Col. (4) is best formed by putting $\gamma \log_{10} e$ on the plate of the arithmometer, multiplying it by $1/x$, obtained from a table of reciprocals and reading off the result negatively.

The point to be borne in mind in drawing a curve of this type is that as the mode and origin are not at the same place, care must be taken to give the maximum ordinate its right position and magnitude (cf. Type IV).

The graduated figures agree fairly closely with the original statistics below the 90–94 group, but are unsuitable for that and the two later groups. The reason is that Type IV, having an unlimited range, should be used. The particular case was chosen partly because an example in which μ_3 is negative is rather more awkward than when μ_3 is positive. In such cases it is a good check to imagine the statistics written in inverse order (in this case 4, 18, 53, etc.), and so avoid the negative signs.

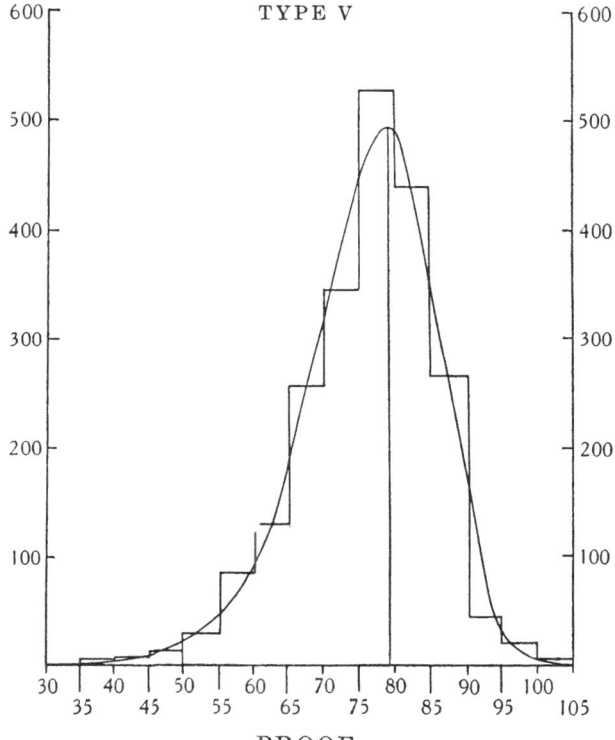

TYPE V

PROOF

Putting $\gamma/x = z$ in $y = y_0 e^{-\gamma/x} x^{-p}$, and integrating from 0 to ∞, we have
$$N = y_0 \gamma^{1-p} \, \Gamma(p-1)$$
or
$$y_0 = \frac{N \gamma^{p-1}}{\Gamma(p-1)}$$

Using the same substitution, the nth moment about the origin is
$$\mu_n' = \frac{y_0}{N} \gamma^{n-p+1} \int_0^\infty e^{-z} z^{p-n} dz$$
$$= \frac{y_0}{N} \gamma^{n-p+1} \Gamma(p-n-1)$$
$$= \gamma^n \frac{\Gamma(p-n-1)}{\Gamma(p-1)}$$

This gives
$$\mu_1' = \frac{\gamma}{p-2}$$

(99)

which is the distance between the mean and origin,

$$\mu_2' = \frac{\gamma^2}{(p-2)(p-3)}$$

$$\mu_3' = \frac{\gamma^3}{(p-2)(p-3)(p-4)}$$

Transferring the moments to the centroid vertical,

$$\mu_2 = \frac{\gamma^2}{(p-2)^2(p-3)}$$

and
$$\mu_3 = \frac{4\gamma^3}{(p-2)^3(p-3)(p-4)}$$

∴
$$\beta_1 = \frac{\mu_3^2}{\mu_2^3} = \frac{16(p-3)}{(p-4)^2} = \frac{16}{p-4} + \frac{16}{(p-4)^2}$$

and
$$(p-4)^2 - \frac{16}{\beta_1}(p-4) - \frac{16}{\beta_1} = 0$$

$p-4$ will have to be taken as the positive root of the equation, or γ, which from the above equations is given by

$$(p-2)\sqrt{\{\mu_2(p-3)\}},$$

will be imaginary.

Since the tangent to the curve at the top of the maximum ordinate is parallel to the axis of x, the position of the mode is such that dy/dx is zero there, i.e. $y_0 x^{-p-1} e^{-\gamma/x} \left\{ -p + \frac{\gamma}{x} \right\}$ is zero. $x = 0$ and $x = \infty$ give the cases in which the curve touches the axis of x, and the other case, the one required, is when $p - \frac{\gamma}{x} = 0$,

or $x = \frac{\gamma}{p}$, i.e. the mode is $\frac{\gamma}{p}$ from the origin.

Uncommon Frequency Types

Up to the present we have dealt with common types of frequency-curves, but in the course of statistical work a distribution is sometimes found which appears different in its algebraic form from the usual types, but can nevertheless be described accurately by those types. An example which will give an indication of the kind of case we have in mind is a distribution arising from recording the number of sequences in coin-tossing or dice-throwing experiments: the distribution is a geometric progression and this, a well-known result in probability, is a special case of Type III if $p = 0$, for we then obtain the exponential $e^{-\gamma x}$ which gives the series we want. Certain limiting cases of Types I, II and VI give straight lines, curves starting with an infinite and ending with a finite ordinate, two separated blocks of frequency, and curves starting at a finite ordinate and ending at zero either at a finite point or at infinity: among these last is, of course, the exponential to which we have already referred.

Before turning to the expressions for these new types it may be useful to give a table of various peculiar distributions that have been obtained from insurance and other material.

Examples of uncommon Frequency Types

469	45	119	4,165	33	68
186	38	100	2,028	53	24
166	46	86	982	65	17
134	53	75	480	81	14
122	43	61	266	101	12
112	38	50	132	131	11
...	49	39	71	186	10
...	41	27	36	350	10
...	44	22	17	...	10
...	52	12	9	...	11
...	...	3	2	...	12
...	...	...	1	...	20
...	...	...	1	...	...
...	...	...	1	...	...
...	...	...	1	...	...
1,189	449	594	8,192	1,000	219

The table includes (col. 6) areas of a U-shaped curve which is rare; in fact, I have not succeeded in finding a suitable distribution of this shape among actuarial statistics, but such a distribution might occur among terminations (including withdrawals) in term policies of ten years, say, or similar endowment assurances.

We may now deal with these cases, but we shall discuss them in less detail than the more important types.

Type VIII

$$y = y_0 \left(1 + \frac{x}{a}\right)^{-m}$$

Range from an infinite ordinate at $-a$ to a finite ordinate, y_0, at 0.

m is found from the solution of

$$m^3(4 - \beta_1) + m^2(9\beta_1 - 12) - 24\beta_1 m + 16\beta_1 = 0$$

and must be neither < 0 nor > 1.

$$a = \pm \sigma(2 - m)\sqrt{\frac{3 - m}{1 - m}}$$

$$y_0 = N(1 - m)/a$$

The distance of the mean from $x = -a$ is $a(1 - m)/(2 - m)$; and from $x = 0$ is $-a/(2 - m)$. When μ_3 is positive a is negative.

If we use the form with origin at mean (see Table VI, facing p. 51)

$$A = a(1 - m)/(2 - m) \quad \text{and} \quad y_e = N(1 - m)(2 - m)^m/a(1 - m)^m.$$

The curve is a special case of Type I when m_2 is zero, that is, when

$$r - 2 = r(r + 2)\sqrt{\{\beta_1/[\beta_1(r + 2)^2 + 16(r + 1)]\}}$$

where $\quad r = 6(\beta_2 - \beta_1 - 1)/(6 + 3\beta_1 - 2\beta_2)$

Thus the test for the suitability of the curve is that

$$\frac{(4\beta_2 - 3\beta_1)(10\beta_2 - 12\beta_1 - 18)^2 - \beta_1(\beta_2 + 3)^2(8\beta_2 - 9\beta_1 - 12)}{(3\beta_1 - 2\beta_2 + 6)\{\beta_1(\beta_2 + 3)^2 + 4(4\beta_2 - 3\beta_1)(3\beta_1 - 2\beta_2 + 6)\}}$$

or λ, say, is zero.

The criteria for Type VIII can be reduced to (1) special case of Type I, (2) $\lambda = 0$, (3) $5\beta_2 - 6\beta_1 - 9$ is negative. It may be added that $24\beta_2 - 27\beta_1 - 38$ is small; theoretically positive.

If $\beta_1 = 0$ an interesting special case arises, in which $m = 0$, and the curve becomes a horizontal line, which is also the limit of Types IX and XII.

The solution of the cubic for m gives trouble. m can also be found from $m = -2(5\beta_2 - 6\beta_1 - 9)/(3\beta_1 - 2\beta_2 + 6)$, and though this involves β_2 it should theoretically give the same value of m as the cubic. As the criterion is not exactly reached in practice the two results differ, and it seems preferable to find m from the cubic by using

$$m = \frac{24\beta_1 - \sqrt{\{24^2\beta_1^2 - 64\beta_1[m'(4 - \beta_1) + 9\beta_1 - 12]\}}}{2\{m'(4 - \beta_1) + 9\beta_1 - 12\}}$$

where m' is found from the expression in β_1 and β_2 given above or by some other trial method.

An alternative is to find from the criteria or from the diagram in *Tables for Statisticians* the value of β_2 which is the consequence of the particular value of β_1 when a Type VIII curve occurs, and use this theoretical value in finding m instead of the β_2 given by the actual statistics.

Example

Frequency	Graduation (1)	Graduation (2)
469	437	436
186	222	209
166	165	161
134	136	141
122	120	127
112	109	115
1,189	1,189	1,189

The mean is $\cdot 65518$ of an interval after the centre of the 186 group. The constants were

$$\mu_2 = 2\cdot986 \qquad\qquad \beta_1 = \cdot408$$
$$\mu_3 = 3\cdot295 \qquad\qquad \beta_2 = 2\cdot047$$
$$\mu_4 = 18\cdot252 \qquad\qquad \lambda = -\cdot05$$

$5\beta_2 - 6\beta_1 - 9$ negative. Hence Type VIII can be used.

$$m = \quad \cdot 500$$

$$a = -5 \cdot 797$$

$$y_0 = 102 \cdot 6$$

The curve runs from $\cdot 277$ before the middle of the first to $\cdot 02$ after the end of the last group. The graduation is shown (No. 1) above.

The areas can be calculated by the expression

$$y_{-r}(a-r)/(1-m)$$

which gives the area of the remainder from $-r$ to $-a$. In the particular case the range could be fixed at 6 as the data related to six months' experience of maturities among endowment assurances, and remembering that the mean is

$$a(1-m)/(2-m)$$

we found

$$m = \cdot 439 \qquad y_0 = 111 \cdot 1$$

The areas resulting are given in graduation (2). The following table gives the calculation of the areas in this case. The equation to the curve is

$$y = 111 \cdot 1 \left(1 - \frac{x}{6}\right)^{-\cdot 439}$$

with range from 0 to 6, a being negative because μ_3 is positive.

x	$1-x/6$	Colog (2)	(3) $\times m$	(4) $+\log 111 \cdot 1$ $= \log y_x$	Log $\dfrac{y_x}{1-m}$	Antilog (6)	Remainder of range	(7) $\times$ (8)	Area required
(1)	(2)	(3)	(4)	(5)	(6)	(7)	(8)	(9)	(10)
0	...	...	...	...	...	...	6	1,189	115
1	·8333	·0792	·0348	2·0804	2·3318	214·7	5	1,074	127
2	·6667	·1761	·0774	2·1230	2·3744	236·8	4	947	141
3	·5000	·3010	·1323	2·1779	2·4293	268·7	3	806	161
4	·3333	·4815	·2118	2·2574	2·5088	322·7	2	645	209
5	·1667	·7781	·3419	2·3875	2·6389	435·5	1	436	436

Some of the columns can be dispensed with: they are shown in detail to make the method clear.

Both graduations are reasonably close to the facts.

An example of the limiting case will be found in the following statistics:

No.	Frequency	Graduation	Theoretical
1	45	42	45
2	38	45	45
3	46	45	45
4	53	45	45
5	43	45	45
6	38	45	45
7	49	45	45
8	41	45	45
9	44	45	45
10	52	48	45
...	449	450	450

The mean is $\cdot 57$ after the middle of the 5 group; the moments are

$$\mu_2 = 8\cdot374 \qquad\qquad \beta_1 = \cdot011$$
$$\mu_3 = \cdot026 \qquad\qquad \beta_2 = 1\cdot78$$
$$\mu_4 = 124\cdot46$$

Hence
$$y = \frac{449}{2\sqrt{(3\mu_2)}} = 44\cdot8$$

The range is from $\cdot57$ to $10\cdot43$.

The series was found by summing in tens the last figure of Carlisle $3\frac{1}{2}$ per cent. Table of A_x and the mean should be at $5\cdot5$ theoretically instead of $5\cdot57$ and y should be 45. The range should be $\cdot5$ to $10\cdot5$. The example is interesting as showing how the Pearson-curves graduate in an extreme case. The "graduation" and theoretical results are shown. In the "graduation" decimals have been neglected.

Type IX

$$y = y_0\left(1+\frac{x}{a}\right)^m$$

Range from $x = -a$ where $y = 0$ to $x = 0$ where $y = y_0$

$$a = \pm\sigma(m+2)\Big/\sqrt{\left(\frac{m+3}{m+1}\right)}$$

m is found by solving

$$m^3(\beta_1 - 4) + m^2(9\beta_1 - 12) + 24m\beta_1 + 16\beta_1 = 0$$
$$y = N(m + 1)/a$$

The distance of the mean from $x = -a$ is $a(m + 1)/(m + 2)$ and from $x = 0$ is $-a/(m + 2)$.

If we use the form with origin at mean (see Table VI, facing p. 51), $A = (m + 1)a/(m + 2)$ and $y_e = \dfrac{N(m + 1)^{m+1}}{a(m + 2)^m}$

As in Type VIII, the value of m can be found by simplifying the cubic into a quadratic, or by the other method indicated.

The criteria are reached through the same equation as those for Type VIII, and can be reduced to (1) special case of Type I, (2) $\lambda = 0$, (3) $5\beta_2 - 6\beta_1 - 9$ is positive, (4) $2\beta_2 - 3\beta_1 - 6$ is negative.

If $\beta_2 = 2\cdot4$ and $\beta_1 = \cdot32$, the curve becomes a sloping line

$$y = \frac{\sqrt{2}N}{3\sigma}\left(1 + \frac{x}{3\sqrt{2}\sigma}\right)$$

If $\beta_1 = 0$ we reach a horizontal line as the limit, while if $\beta_2 = 9$ and $\beta_1 = 4$, we have the other limit of Type IX, and find the exponential series (Type X).

Example

Duration	Exposed to risk in annuity experience	Type IX	Frequency line
0	119	118	108
1	100	98	97
2	86	85	86
3	75	74	76
4	61	63	65
5	50	52	54
6	39	41	43
7	27	30	32
8	22	20	22
9	12	11	11
10	3	2	0
...	594	594	594

The mean is at 2·909 assuming the exposed to risk to be an ordinate at the duration or an area from $n - \frac{1}{2}$ to $n + \frac{1}{2}$.

$$\mu_2 = \quad 6\cdot27 \qquad\qquad \beta_1 = \quad \cdot490$$
$$\mu_3 = \quad 10\cdot99 \qquad\qquad \beta_2 = 2\cdot606$$
$$\mu_4 = 102\cdot50$$

$5\beta_2 - 6\beta_1 - 9$ is positive $\qquad 2\beta_2 - 3\beta_1 - 6$ is negative

The curve is not far from Type IX, and if β_1 had been ·32 and β_2 had been 2·4, we should have reached a straight line

$$y = 111\cdot8\left(1 - \frac{x}{10\cdot63}\right)$$

with range from $-\cdot6$ to $10\cdot0$ and obtained the graduation shown. The whole area $-\cdot6$ to $+\cdot5$ is taken as the frequency for duration 0. Using Type IX, the following constants are reached:

$$m = 1\cdot123, \qquad a = -10\cdot913, \qquad y_0 = 115\cdot54$$

The curve runs from $-\cdot586$ to $10\cdot3275$. The 118 in the first group has been taken as the area from $-\cdot586$ to $+\cdot5$. Theoretically there cannot be an exposure before duration $-\cdot5$, but as we are merely giving an example of fitting a curve to a series of numbers this need not concern us. The difficulty could be met by fitting a system of ordinates or by assuming a starting point for the curve.

If m happens to be less than unity the shape of the curve is somewhat different, e.g. if $y = 100\left(1 + \dfrac{x}{10}\right)^{\cdot25}$ we have the following ordinates:

100, 98, 95, 91, 88, 84, 79, 74, 67, 56, 0

The actual deaths in a select mortality experience may take this form, but the shape of the curve will be less flat at the start, e.g. in the American Medico-Actuarial experience 1913 age group 30–34.

Type X

$$y = \frac{N}{\sigma} e^{-x/\sigma}$$

Range from 0 to ∞.

Distance of origin from the mean is σ.

The ordinate at the mean (y_e) is $N/e\sigma$.

The curve is a special case of Type III when $\gamma a = 0$, that is when $\beta_1 = 4$.

The condition for Type III is given by $2\beta_2 = 6 + 3\beta_1$. Hence the exponential form is given by $\beta_1 = 4$ and $\beta_2 = 9$. The curve is also the limit of Types IX and XI.

Example

	Frequency	Graduation	Theoretical
1	4,165	4,132	4,096
2	2,028	2,016	2,048
3	982	1,015	1,024
4	480	511	512
5	266	257	256
6	132	130	128
7	71	65	64
8	36	33	32
9	17	17	16
10	9	8	8
11	2	4	4
12	1	2	2
13	1	1	1
14	1	1	1
15	1	...	...
...	8,192	8,192	8,192

The unadjusted mean is 2·0087.

$$\mu_2 = 2·045 \qquad \beta_1 = 4·629$$
$$\mu_3 = 6·290 \qquad \beta_2 = 9·502$$
$$\mu_4 = 39·720 \qquad \sigma = 1·43$$

When the curve is an exponential the moments and mean require adjustment, but the Sheppard high contact adjustments are, of course, unsuitable. If the curve starts at the beginning of the first group, I think that the mean is overstated when μ_3 is positive by $1/12\sigma$ approximately, and the second moment about the true mean is understated by $\frac{1}{12}$ approximately.* Making use of the adjustments the mean is now 1·934, μ_2 is 2·123, and σ is 1·457.

The statistics relate to sequences in coin-tossing and the theoretical figures are added. In the statistics as published the sequences of 11, 12, etc. were 2, 1, 0, 1, 0, 2. Strictly speaking we are dealing with a system of ordinates; I made the calculation as a series of areas in order to introduce the adjustment of moments. In calculating the graduated areas of the curve it is useful to remember that the area from a to b is $(y_a - y_b)\sigma$.

It is interesting to notice how the "graduation" keeps closer to the frequency than the theoretical result.

I give as a second example the following series based on cricket scores known to start at the beginning of the first group:

Score ...	0–19	20–	40–	60–	80–	100–	120–	140–	160–
Series ...	64	34	18	9	6	3	3	0	0
Graduation	64	34	18	10	5	3	1	1	1

The ratio of each term to the preceding is ·54, and the graduation is almost exact. Owing, however, to the 3 at the group 120, the moments give a criterion considerably removed from the theoretical $\beta_1 = 4$, $\beta_2 = 9$.

If we had assumed the start of the curve in the previous example, we should have reproduced the theoretical result almost exactly.

---

* See, however, general discussion in Appendix I.

Type XI

$$y = y_0 x^{-m}$$

Range from $x = b$ where $y = y_0 b^{-m}$ to $x = \infty$ where $y = 0$.
m is found from

$$m^3(4 - \beta_1) + m^2(9\beta_1 - 12) - 24\beta_1 m + 16\beta_1 = 0$$

$$b = \pm \sigma(m - 2) \sqrt{\frac{m - 3}{m - 1}}$$

$$y_0 = N b^{m-1}(m - 1)$$

The distance of mean from origin is $b(m - 1)/(m - 2)$.

If we use the form with origin at mean (see Table VI, facing p. 51), $A = b(m - 1)/(m - 2)$ and

$$y_e = \frac{N}{b} \cdot \frac{(m - 2)^m}{(m - 1)^{m-1}}$$

As in Type VIII, m can be found by simplifying the cubic into a quadratic or by the other method indicated.

m may have any value from 5 to ∞, but in practice its value is not less than 9.

The curve is a special case of Type VI when $q_2 = 0$.

The criteria can be expressed as (1) special case of Type VI, (2) $\lambda = 0$, (3) $2\beta_2 - 3\beta_1 - 6$ is positive.

Example

Duration	Withdrawals	Graduation by XI
0	165	183
1	65	53·9
2	23	32·6
3	32	20·0
4	13	12·4
5	8	7·6
6	1	4·9
7	6	3·1
8	3	1·9
9	3	1·2
10	1	·8
11	3	1·6
...	323	323

I have not come across a distribution really represented by this type, but I give an unsuccessful attempt to apply it to a series of withdrawals. The constants were

$$\beta_1 = 4\cdot97 \qquad\qquad b = 57\cdot14$$
$$m = 29\cdot69 \qquad\qquad \log y_0 = 54\cdot3563$$

Distance of mean from origin $= 59\cdot205$

In calculating areas we use $y_0 a^{-(m-1)}/(m-1)$ as the area from a to ∞.

Twisted J-shaped Curve

As pointed out in the notes on Type I (p. 59), we obtain an interesting curve when both m_1 and m_2 are numerically less than unity and one of them is negative. It arises when

$$\beta_2 > 1\cdot5 + 1\cdot125\beta_1$$

and when $\qquad\qquad \beta_2 < 2 + 1\cdot25\beta_1$

as can be seen by remembering that the sum of the values of the m's must lie between 1 and -1 or r lies between 3 and 1. A special case has been discussed as a transition type (No. XII) when

$$y = y_0 \left(\frac{\sigma\{\sqrt{(3+\beta_1)} + \sqrt{\beta_1}\} + x}{\sigma\{\sqrt{(3+\beta_1)} - \sqrt{\beta_1}\} - x} \right)^{\sqrt{\{\beta_1/(3+\beta_1)\}}}$$

Range from $x = \sigma(\sqrt{(3+\beta_1)} - \sqrt{\beta_1})$ to $x = -\sigma(\sqrt{(3+\beta_1)} + \sqrt{\beta_1})$.
The origin is at the mean.

$$y_0 = \frac{N}{b\,\Gamma(m+1)\,\Gamma(1-m)}$$

where $m = \sqrt{\dfrac{\beta_1}{3+\beta_1}}$ and $b = 2\sigma\sqrt{(3+\beta_1)}$.

When μ_3 is positive, the negative sign is taken for the square roots.

The limit of the curve when $\beta_1 = 0$ is a horizontal line.
The criterion is $5\beta_2 - 6\beta_1 - 9 = 0$.

Example

Frequency	Graduation	Ordinates
...	2	18·5
33	31	31·6
...	...	40·6
53	49	49·3
...	...	57·0
65	65	65·2
...	...	73·4
81	83	82·4
...	...	92·3
101	103	103·3
...	...	114·6
131	134	132·9
...	...	155·5
186	191	186·0
...	...	244·2
350	342	405·4
1,000	1,000	...

The mean is ·051 after the centre of the 131 group. The constants are

$$\mu_2 = \quad 4\cdot266 \qquad\qquad \beta_1 = \quad \cdot761$$
$$\mu_3 = -7\cdot688 \qquad\qquad \beta_2 = \quad 2\cdot646$$
$$\mu_4 = \quad 48\cdot154 \qquad 5\beta_2 - 6\beta_1 - 9 = -\cdot368$$
$$y = 87\cdot2\{(5\cdot808 + x)/(2\cdot204 - x)\}^{\cdot45}$$

In addition to the graduation a number of equidistant ordinates is given. They show that the curve rises abruptly, then less abruptly and then again more abruptly. The withdrawals in select tables are sometimes of this shape (e.g. Japanese experience, 1910, age 52, females). A somewhat similar twist occurs in a population curve.

U-*shaped Curve*

This shape arises in Type I when m_1 and m_2 (or m in Type II) are negative. There are difficulties in fitting it to statistics because it is awkward to adjust the rough moments. The figures given in the table of examples (p. 101, col. 6) were found by calculating the areas of the curve

$$y = 10\left(1+\frac{x}{8}\right)^{-7}\left(1-\frac{x}{4}\right)^{-\cdot35}$$

The limit of the U-curve is two separate blocks of frequency at the ends of the range. This limit is reached when

$$\beta_2-\beta_1-1 = 0.$$

Some of the curves with which we have dealt are rare and in practical curve-fitting may be avoided, for they depend on certain definite values of β_1 and β_2, and the chance of reaching these exact values is negligible. In other words, if the object of fitting a curve in any particular case is to obtain the closest agreement between the actual figures given and the graduated figures, then the main Types (I, IV and VI) are all that are necessary, for the other types being transition types and depending on specific values of β_1 and β_2 need not arise. If, however, our object is to study probability in a wider sense, the transition types are of importance and they may, of course, be properly used when the values of the β's only differ from those indicated by the criteria to a small extent. This "small extent" means (as we shall see later) within the limit suggested by the standard errors of the β's.

ADDITIONAL EXAMPLES

1. Up to the present we have merely considered examples with a view to illustrating the various types of frequency-curves, but it seems advisable to consider one or two practical examples which may help to show the range of applicability of the curves in actuarial work, and give an opportunity of noticing a few difficulties which may arise in applying them.

The function with which actuaries generally wish to deal in practical work is not an exposed to risk or series of deaths or withdrawals, but the ratio between the deaths and the exposed; that is, with the rates of mortality, sickness, marriage, and withdrawal. An actuary studying frequency-curves may therefore naturally ask whether any of these rates can be graduated by means of the curves we have examined, and, if they fail, must they be put aside for some other method? Now the first point to be considered is whether these rates are frequency distributions; if they are not, the use of the frequency-curve is empirical. A rate of mortality gives the proportion of people at each age who die, and if we imagine 1,000 persons exposed to risk at each integral age, the number of deaths would be 1,000 times the rate of mortality, and this seems to show that it is possible to consider the rate of mortality as a distribution, though it is one that could hardly arise in actual experience. It is impossible to describe the rates of mortality or sickness by a single frequency-curve. On the other hand, the rates of marriage are certainly much like frequency-curves, and the rates of withdrawal, whether regarded according to age or duration, might take a form like our example in Type III. There are, however, practical objections to the direct operation on rates, even apart from the very exaggerated idea of frequency distributions in which it is necessary to indulge. The numbers exposed to risk at the end of any table become small, and a single death or marriage there gives a very large rate, while at several ages near there may be a zero rate shown by the ungraduated data. This is extremely awkward, as it tends to make the ratios dealt with far rougher in application than the actual observations are in fact, and we are forced to group the material before using it, which introduces an arbitrary practice which it is well to avoid so far as possible. It must not, of course, be inferred that a small number of say fifty or one hundred deaths must necessarily be grouped according to each year of age, but that even if there are two or three thousand the roughnesses introduced by the use of rates influence the result considerably. Graduating rates means that an equal weight is given to each rate of mortality which is far from the weight indicated by the exposed to risk.

2. It will be useful to consider a case bearing out these

objections and then deal with a practical method of overcoming them. The statistics to be considered have been taken from a paper by M. Mackenzie Lees "On Rates of Mortality and Marriage among daughters of Peers and Heirs Apparent, etc." (*Trans. Fac. Actu.* I, 276), and may be summarised as on p. 117. The moments were calculated by the Summation Method, and were found, about the mean 28·77191, to be

$$\mu_2 = \quad 63\cdot2092 \qquad \beta_1 = 1\cdot557153$$
$$\mu_3 = \quad 627\cdot101 \qquad \beta_2 = 4\cdot781321$$
$$\mu_4 = 19,103\cdot3$$

The criterion was $\kappa = -1\cdot5$, but as I had neglected the rate ·0089 at 71 in calculating the moments, I used Type III. The inclusion of the rate at that age would have lengthened the curve and considerably increased the arithmetical value of the criterion. Moreover $2\beta_2$ approximates to $6 + 3\beta_1$.

The constants for Type III were

$$\gamma = \quad \cdot201592 \qquad a = \quad 7\cdot78189$$
$$p = 1\cdot56881 \qquad \text{Mode} = 23\cdot81128$$

The curve starts, therefore, at age 16·02939.

$$y_0 = 890\cdot05.$$

The rates resulting from this graduation are given in the table, and while they tend to show that the distribution of rates of marriage is closely allied to a frequency-curve, they do not give a satisfactory graduation, and the failure is due almost entirely to the objections mentioned above. If we were examining the algebraic form taken by rates of marriage, we should begin by work on population data where the roughness of material is avoided by the large numbers of individuals dealt with; as, however, we are seeking for a graduation, we must see how these objections, which of course apply to some extent to any method of graduation, can be overcome. It has been remarked that the cause of the difficulty is that incorrect weights are given to the items used, and the most obvious suggestion is that the actual exposed and marriages should be graduated separately. This, however, entails a large amount of additional work and seems to overlook the fact that deviations in the exposed to risk and the marriages are not independent. A shorter method can be used which avoids both the double graduation and the error just indicated. This method consists of using a series allied to the exposed, and treating it as a hypothetical exposed to risk from which a new series of marriages can be calculated. The advantages are that we have only to make one graduation, and the weights of the various parts of the table are given approximately. In a similar way q_x can be graduated, and in this connection it may be remarked that as the exposed to risk is generally capable of being represented by a frequency-curve, it is natural to suggest that the hypothetical exposed might be taken as the simplest form assumed by such curves (normal curve); this is also convenient because the ordinates for such curves have been tabulated.

3. The hypothetical exposed can be fixed by trial or from the values of the exposed. The column E'_x in the table given on p. 117 is taken from Sheppard's Tables in *Tables for Statisticians*, x being taken as 3·06, 3·084, 3·108, 3·132, etc., and the entries were multiplied by 10^6. $M'_x = E'_x \times m_x$ was then formed and graduated. The following values were obtained for the M'_x series:

Mean =	24·85779	$\beta_1 =$	1·40775
$\mu_2 =$	29·5006	$\beta_2 =$	5·01114
$\mu_3 =$	190·112	$\kappa =$	$-7·102$
$\mu_4 =$	4,361·12		

As κ is large, Type III was used, and

$\gamma =$	·310350	$y_0 =$	192·625
$p =$	1·841405	Mode =	21·63562
$a =$	5·933325		

The curve was then worked out and the rates of marriage in the final column were obtained by dividing M' by E'. They agree closely with the ungraduated figures.

Marriage Rates of Spinsters

Age x	Exposed to risk E	No. of marriages M_x	Rate of marriage m_x	Rate of marriage graduated by frequency-curve	Hypo-thetical exposed E'_x	No. of marriages M'_x	Rate of marriage graduated
15	3,658	3	·0008	...	3,695	3	...
16	3,603	8	·0022	·0027	3,433	7	·0018
17	3,528·5	49	·0139	·0132	3,187	44	·0157
18	3,393·5	114	·0336	·0332	2,957	99	·0350
19	3,187	176	·0552	·0517	2,742	151	·0541
20	2,945	219	·0744	·0667	2,541	189	·0695
21	2,688·5	192	·0714	·0776	2,354	168	·0809
22	2,443	211	·0864	·0846	2,179	188	·0880
23	2,187	212	·0969	·0881	2,016	196	·0917
24	1,956	194	·0992	·0889	1,864	185	·0920
25	1,758	146	·0831	·0875	1,723	143	·0901
26	1,583·5	137	·0865	·0845	1,591	138	·0861
27	1,417	121	·0854	·0803	1,469	126	·0812
28	1,270·5	105	·0826	·0753	1,355	112	·0754
29	1,148·5	75	·0653	·0698	1,249	82	·0693
30	1,068	60	·0562	·0640	1,151	65	·0631
31	984	64	·0650	·0583	1,061	69	·0569
32	904·5	41	·0453	·0528	976	43	·0508
33	848·5	30	·0354	·0475	897	32	·0452
34	802	39	·0486	·0425	825	40	·0400
35	752	20	·0266	·0378	758	20	·0352
36	711	25	·0352	·0335	696	25	·0309
37	672·5	18	·0268	·0295	639	17	·0270
38	638	11	·0172	·0260	586	10	·0235
39	612·5	14	·0229	·0228	537	12	·0205
40	586·5	15	·0256	·0199	492	12	·0176
41	568·5	9	·0158	·0173	451	7	·0151
42	541·5	6	·0111	·0151	412	5	·0130
43	515	8	·0155	·0131	376	6	·0112
44	491·5	2	·0041	·0113	345	1	·0096
45	476	5	·0105	·0097	315	3	·0082
46	454	5	·0110	·0084	288	3	·0070
47	440·5	2	·0045	·0072	262	1	·0060
48	416	5	·0120	·0062	239	3	·0051
49	395	2	·0051	·0054	218	1	·0044
50	378·5	...	...	·0046	199	...	·0037
51	363·5	...	...	·0039	181	...	·0031
52	348·5	1	·0029	·0034	165	...	·0026
53	335·5	2	·0089	·0029	150	1	·0022
54	317·5	...	...	·0024	139	...	·0019
55	304	...	...	·0020	124	...	·0016
56	291	3	·0103	·0018	...	1	...
57	278·5	1	·0036	·0015	...	...	...
58	261	...	...	·0013	...	...	...
59	248·5	...	...	·0011	...	...	...
60	234·5	...	...	·0009	75·7	...	·0007
61	219·5	1	·0046	·0007	...	...	...
62	209·5	...	...	·0006	...	...	...
63	201·5	...	...	·0005	...	...	...
64	191	...	...	·0004	...	...	...
65	177	...	...	·0004	45·7	...	·0003
66	165·5	...	...	·0003	...	...	...
67	154	...	...	·0002	...	...	...
68	147·5	...	...	·0002	...	...	...
69	135·5	...	...	·0001	...	...	...
70	124·5	...	...	·0001	27·2	...	·0001
71	112·5	1	·0089	...	...	...	...
72	105·5	...	...	...	...	...	...
73	95	...	...	...	...	...	...
74	84·5	...	...	...	...	...	...
75	79	...	...	...	...	...	...

4. A numerical example of the application of the method to the O^(NM(5)) Table may now be given. The normal curve with $\sigma = 10$ and origin at age $52\frac{1}{2}$ was used, and the values were multiplied by q_x with the help of Crelle's Tables.

A part of the work was

$q \times E \times 10^5$	Age	Ordinate from Sheppard's Tables $= E$	Age	$q \times E \times 10^5$
810	52	·3984439	53	801
597	51	·3944793	54	850
644	50	·3866681	55	875

Summing these entries ($q \times E \times 10^5$) in fives, I formed the following:

Age	$q \times E \times 10^5$
20	13
25	70
30	218
35	594
40	1,394
45	2,460
50	3,702
55	4,519
60	4,385
65	3,602
70	2,249
75	1,197
80	461
85	133
90	31
95	5
100	1
	25,034

The abbreviations (use of Crelle's Tables and grouping) were adopted to save labour, and as the figures were required for an example they are sufficiently accurate.

The following values were then found:

$$\text{Mean age} = 59 \cdot 439762$$
$$\mu_2 = 4 \cdot 584327$$
$$\mu_3 = - \cdot 4999871$$
$$\mu_4 = 61 \cdot 17014$$

Type of curve.—No. I.

$$m_1 = \quad 32\cdot81166$$
$$m_2 = \quad 26\cdot57123$$
$$a_1 = \quad 18\cdot78553$$
$$a_2 = \quad 15\cdot21272$$
$$y_0 = 4609\cdot884$$
$$\text{Mode age} = \quad 59\cdot730789$$

(The unit is 5 years of age.)

The ordinates were then calculated for every fifth age, and finding that the curve is not very far removed from the normal curve of error, I interpolated in the second differences of the logarithms of the ordinates for those at the other ages.* A quadrature formula was used for finding areas, and q_x was found by dividing by the hypothetical figure already used for the exposed.

The expected deaths were as follows:

Group	Graduated q_x for central age of group	Actual	Expected	Deviation +	Deviation −
15–19	...	...	1·5	1·5	...
20–	·00643	9	8·9	...	·1
25–	·00731	69	61·0	...	8·0
30–	·00850	205	204·6	...	·4
35–	·00991	369	380·7	11·7	...
40–	·01179	588	575·6	...	12·4
45–	·01452	801	811·4	10·4	...
50–	·01866	1,064	1,063·8	...	·2
55–	·02505	1,399	1,386·6	...	12·4
60–	·03516	1,752	1,773·2	21·2	...
65–	·05118	2,164	2,136·7	...	27·3
70–	·07682	2,216	2,261·2	45·2	...
75–	·11648	1,965	1,925·8	...	39·2
80–	·17462	1,237	1,241·9	4·9	...
85–	·24870	494	514·4	20·4	...
90–	·33286	129	126·0	...	3·0
95–	·43289	18	17·3	...	·7
100–	...	1	1·5	·5	...
		14,480	14,492·1	115·8	103·7
				219·5	

* As $e^{-(x-h)^2/2\sigma^2}$ is the equation to normal curve, the logarithm is $Ax^2 + Bx + C$, say. The criterion shows if the curve is nearly normal.

5. It will be interesting to examine a particular case of the method just described, as it is often required by actuaries.

Defining Makeham's hypothesis as $\operatorname{colog} p_x = A + Bc^x$, we take a normal curve $(y_0 e^{-(x-h)^2/2\sigma^2})$ to represent the exposed and multiply by the values of $\operatorname{colog} p_x$. This means that we assume that the products can be represented by

$$y = (A + Bc^x)y_0 e^{-(x-h)^2/2\sigma^2}$$

$$= Ay_0 e^{-(x-h)^2/2\sigma^2} + By_0 e^{-(x^2-2hx+h^2-2\sigma^2 x \log_e c)/2\sigma^2}$$

$$= Ay_0 e^{-(x-h)^2/2\sigma^2} + HBy_0 e^{-(x^2-2[h+\sigma^2 \log_e c]x+[h+\sigma^2 \log_e c]^2)/2\sigma^2}$$

where $H = e^{\{h^2 + 2\sigma^2 h \log_e c + \sigma^4 (\log_e c)^2 - h^2\}/2\sigma^2} = e^{h \log_e c + \frac{\sigma^2}{2}(\log_e c)^2}$

$$\therefore \qquad y = Ay_0 e^{-(x-h)^2/2\sigma^2} + HBy_0 e^{-(x-t)^2/2\sigma^2} \qquad \ldots\ldots(\text{I})$$

i.e. the sum of two normal curves both having the same standard deviation as the exposed curve and one having the same origin.

The difference between the two means gives $\sigma^2 \log_e c$, so
$$\log_{10} c = \frac{t-h}{\sigma^2}\log_{10} e.$$

The whole solution is made very simple by taking moments about the known origin (age h), for $\displaystyle\int_{-\infty}^{+\infty} xy\,dx$ and $\displaystyle\int_{-\infty}^{+\infty} x^2 y\,dx$ (the first two moments) give

$$(t-h)\,N_2{}^* \quad \text{and} \quad N_1\sigma^2 + N_2\{\sigma^2 + (t-h)^2\}\dagger$$

where $N_1 = Ay_0\sigma\sqrt{(2\pi)}$ and $N_2 = HBy_0\sigma\sqrt{(2\pi)}$.

Dividing the values just given by $N_1 + N_2$ (the total frequency), we obtain, as the first moment about the known origin, $\dfrac{(t-h)\,N_2}{N_1+N_2}$, and, as the second,

$$\frac{N_1\sigma^2 + N_2\sigma^2 + N_2(t-h)^2}{N_1+N_2} = \sigma^2 + (t-h)\,\mu_1'$$

* Remember that the normal curve is symmetrical, so that the odd moments about the mean of such a curve are zero.

† Can be seen at once as the sum of two integrals; $N_1\sigma^2$ gives the second moment of the first normal curve in (I), and $N_2\{\sigma^2 + (t-h)^2\}$ gives the second moment of the second normal curve.

or
$$t - h = \frac{\mu_2' - \sigma^2}{\mu_1'}$$

and
$$N_2 = \frac{\mu_1'(N_1 + N_2)}{t - h}$$

where μ' is written for moments about h.

As stated above

$$\log_{10} c = \frac{t - h}{\sigma^2} \log_{10} e \qquad \ldots\ldots\text{(II)}$$

and if $y_0 = \dfrac{10^k}{\sigma\sqrt{(2\pi)}}$ as is generally convenient, then

$$A = N_1/10^k$$

and
$$B = N_2/(10^k \times H)$$

$$= \frac{N_2}{10^k} \cdot \frac{1}{e^{h\log_e c + \frac{\sigma^2}{2}(\log_e c)^2}}$$

$$= \frac{N_2}{10^k c^h e^{\frac{(t-h)}{2}\log_e c}} \qquad \text{(see equation (II))}$$

$$= \frac{N_2}{10^k c^{\frac{t+h}{2}}}$$

Care is necessary with regard to the value used for y_0, and consequently with regard to A and B. If Sheppard's tables in *Tables for Statisticians* of ordinates (z) be multiplied by, say, 10^5 and used as the exposed to risk, the values of A and B resulting from the work will be $N_1/(10^5\sigma)$ and $N_2/(10^5 H\sigma)$. The reason is that his tables are in terms of standard deviation.

6. If we assume, as Hardy did when graduating the British Offices 1863–1893 experience, that $\log_{10} c$ is known, we only require to calculate one moment which gives us $\dfrac{(t - h) N_2}{N_1 + N_2}$, and this, with the help of equation (II), enables us to complete the solution. If c were obtained for the aggregate table, we should use this result for the select tables.

7. A numerical example with the $O^{NM(5)}$ Table may be of interest. A normal curve with standard deviation 10 and origin $55\frac{1}{2}$ was taken, and the terms multiplied by $\operatorname{colog} p_x$. These were then grouped in fives, and the first two moments calculated about age $55\frac{1}{2}$. One little point should be borne in mind in connection with the grouping; though the centre of the base on which the product $(q_x \times$ exposed$)$ stands is $x + \frac{1}{2}$, the result $(\operatorname{colog} p_x \times$ exposed$)$ is an ordinate at x; the centre point of five ages 20 to 24 is $22\frac{1}{2}$ when q_x is used and 22 when $\operatorname{colog} p_x$ is used.

The figures were

$$N_1 + N_2 = 136387$$

1st moment about $55\frac{1}{2}$ in 5-years unit $= 1\cdot416184$

2nd ,, ,, ,, $= 4\cdot1929354$

Deducting Sheppard's adjustment of $\frac{1}{12}$ from the second moment* and multiplying the first moment and the adjusted second moment by 5 and 25 respectively to make the unit one year instead of 5 years, we have

$$\mu_1' = 7\cdot080920$$
$$\mu_2' = 164\cdot384085$$
then
$$\log(t-h) = \cdot9586889$$
$$t-h = 9\cdot092617$$
$$\log_{10} c = \cdot03948873$$
$$A = \cdot00301749$$
$$B = \cdot00004518782$$
$$\log_{10} B = \bar{5}\cdot6550214$$

q_x was then calculated from the graduated $\operatorname{colog} p_x$ obtained from the values of A, B and c, and the following table of

---

* As we are dealing with the sum of five ordinates in each group and not with an area, we should not, strictly speaking, use Sheppard's adjustment, but should deduct $\cdot08$. The difference is small and the constants have not been re-calculated. The formulae would be $\mu_2 = \nu_2 - \cdot08$ and $\mu_4 = \nu_4 - \cdot48\nu_2 + \cdot02752$ where μ is adjusted and ν unadjusted.

expected deaths was worked out. The values of q_x are given in the table showing the frequency-curve graduation:

Age group	Graduated q_x for central age of group	Expected deaths	Actual deaths	Deviation +	Deviation −
Under 25	...	13·0	9	4·0	...
25–29	·00812	67·0	69	...	2·0
30–	·00882	211·6	205	6·6	...
35–	·00991	380·8	369	11·8	...
40–	·01162	566·9	588	...	21·1
45–	·01431	799·7	801	...	1·3
50–	·01854	1,057·5	1,064	...	6·5
55–	·02517	1,392·7	1,399	...	6·3
60–	·03551	1,790·2	1,752	38·2	...
65–	·05160	2,153·0	2,164	...	11·0
70–	·07639	2,249·3	2,216	33·3	...
75–	·11415	1,888·7	1,965	...	76·3
80–	·17053	1,213·6	1,237	...	23·4
85–	·23352	519·1	494	25·1	...
90–	·36484	136·6	129	7·6	...
95–	...	20·6	19	1·6	...
...	...	14,460·3	14,480	128·2	147·9

276·1

This result is very like that given by the late Sir G. F. Hardy, but avoids having to obtain c by trial. Hardy's expected and actual deaths balance better than the above, but I do not think the rates have been understated systematically; the 75–79 group accounts for the disagreement. The total deviation is less than Hardy's.

8. Another possible application of frequency-curves to life assurance and mortality statistics was discussed recently. The exposed to risk or the amount of the sums assured or premiums at each age can usually be graduated by a frequency-curve. When an actuary values the liabilities of an insurance company he works, in effect, on the proportion of the business that survives to each age in successive years according to the mortality table assumed in the valuation. If the proportion, at age x, that survives n years by a given table of mortality is $_np_x$ and if E_x is the amount of sums assured, say, on the books at age x, then the amount of sums assured surviving after n

years is $E_x \cdot {}_n p_x$. For diverse mortality tables, various values of n and a fairly wide range of frequency-curves assumed for E_x, we again reach a frequency-curve as an approximation to the distribution of $E_x \cdot {}_n p_x$ in terms of x. Several statistical examples have been given* and the reader who wishes to examine other examples than those given in this book may refer to them or to such a large collection of examples as those given by K. Pearson and A. Lee for Barometric Heights.†

9. If we know the range of a curve we need not even with Type I find as many as four moments, for the equations on p. 64, giving the moments about the start of the curve, afford a simple solution. We have

$$\mu_1' = \frac{b(m_1 + 1)}{m_1 + m_2 + 2} \quad \text{and} \quad \mu_2' = \frac{b^2(m_1 + 1)(m_1 + 2)}{(m_1 + m_2 + 2)(m_1 + m_2 + 3)}$$

and writing

$$\gamma_1 = \frac{\mu_1'}{b} \quad \text{and} \quad \gamma_2 = \frac{\mu_2'}{\mu_1' b}$$

we have

$$m_1 + 1 = \frac{\gamma_1(\gamma_2 - 1)}{\gamma_1 - \gamma_2}$$

and

$$m_2 + 1 = \frac{(\gamma_2 - 1)(1 - \gamma_1)}{\gamma_1 - \gamma_2}$$

where μ' is written for a moment about the start of the curve.

10. If, however, we can only fix by general considerations the start of the curve, the following solution depending on three moments is of use.

Writing

$$\lambda_2 = \frac{\mu_2'}{\mu_1'^2} \quad \text{and} \quad \lambda_3 = \frac{\mu_3'}{\mu_2' \mu_1'}$$

the values of the constants in the equation to the curve are given by

$$m_1 + 1 = \frac{2(\lambda_2 - \lambda_3)}{2\lambda_3 - \lambda_2 - \lambda_2 \lambda_3}$$

* J. Inst. Actu. LXV, 1.   † Philos. Trans. A, CXC, 423.

$$m_2 + 1 = \frac{2(\lambda_2 - \lambda_3)(\lambda_3 - 1)(1 - \lambda_2)}{(2\lambda_3 - \lambda_2 - \lambda_2\lambda_3)(1 + \lambda_3 - 2\lambda_2)}$$

$$b = \mu_1' \frac{m_1 + m_2 + 2}{m_1 + 1}$$

and $\qquad a_1/a_2 = m_1/m_2$

11. We may return to Type I for an example of the method of §9 where we will assume that the curve starts at age 17·5 and has a range of 15·5 units Considering the line for age 22 in the table on p. 60 we see that 4·175 and 14·634 give S_2 and S_3, excluding the first group, and the moments about age 17 are then found to be 4·175 and 25·093; transferring to 17·5, we have 4·075 and 24·268; adding the moments for the first group, ·034 × $\frac{1}{5}$ and ·034 × $(\frac{1}{5})^2$ respectively, $\mu_1' = 4\cdot0818$ and

$$\mu_2' = 24\cdot26936$$

Hence
$$m_1 = \cdot3498 \qquad a_1 = 1\cdot735$$
$$m_2 = 2\cdot7758 \qquad a_2 = 13\cdot765$$
$$y_0 = 154\cdot2$$

and the mode is $17\cdot5 + 1\cdot735 \times 5 = 26\cdot175$

From these values the graduated figures for the first four groups are 37, 140, 152, 143.

12. We can use this method with Example I, p. 7. With no moment adjustment we obtain

$$y = \cdot89082 x^{-\cdot629685} (25\cdot49729 - x)^{1\cdot624275}$$

with origin at 1·02897 where the curve starts.* This shows that the first term (308) should be assumed to be at less than unit distance before the second term (200). Recalculating the first two moments by assuming the distance between the first two terms to be ·75 and fixing the range at 25 and start at 1·1, I found

$$y = \cdot89055 x^{-\cdot61830} (25 - x)^{1\cdot63900}.$$

---

* The graduation on p. 125 in 3rd edition is not reached by this curve. I must have used an adjustment similar to that now given but cannot recall it.

The graduation by this curve is shown in the following table:

Duration	Withdrawals	Graduated by Type I curve
1	308	315
2	200	192
3	118	100
4	69	74
5	59	57
6	44	46
7	29	38
8	28	31
9	26	26
10	21	22
11	18	19
12	18	16
13	12	13
14	11	11
15	5	9
16	11	7
17	7	6
18	6	5
19	1	4
20	3	3
21	1	2
22	3	2
23	2	1
24	...	1
	1,000	1,000

13. The calculation of the graduated area of the first group may present a difficulty, as a quadrature formula cannot be applied, and the following method gives the best way of obtaining a correct value

$$\int_0^x y_0' x^{m_1}(b-x)^{m_2} dx$$

$$= \int_0^x y_0' x^{m_1}\left(b^{m_2} - m_2 b^{m_2-1} x + \frac{m_2(m_2-1)}{2} b^{m_2-2} x^2 - \ldots\right) dx$$

$$= y_0' x^{m_1+1} b^{m_2}\left(\frac{1}{m_1+1} - \frac{m_2 x}{b(m_1+2)} + \ldots\right)$$

which is a rapidly convergent series when x is small In the preceding example, where x is $1\cdot5 - 1\cdot02897 = \cdot47103$, the

(126)

second term barely affects the result. y_0' must be calculated by the formula

$$\frac{N}{b^{m_1+m_2+1}} \cdot \frac{\Gamma(r)}{\Gamma(m_1+1)\,\Gamma(m_2+1)}$$

The expression for finding the area of the first group in Type III curves is

$$\int_0^x y_0' e^{-\gamma x} x^p \, dx = y_0' x^{p+1} \left(\frac{1}{p+1} - \frac{\gamma x}{p+2} + \dots \right)$$

where $y_0' = N\gamma^{p+1}/\Gamma(p+1)$.

CHAPTER VI

COMPARISON OF VARIOUS SYSTEMS
OF CURVES

1. In the previous chapter we dealt with Pearson's system of frequency-curves, but other methods have been used to describe frequency distributions. We have already seen that Pearson's system of curves describes the facts that have been collected about a variety of subjects connected with chance. A system is useless if it does not give approximately the distributions that actually occur. The binomial series is justified from this point of view as a description of the number of times events happen, because we have found from experience that the numbers given by it are realised approximately by trial. When we consider the matter we are almost compelled to admit that the real justification of any theory of probability is that events happen in the way such a theory leads one to expect, and if we wish to compare the systems of frequency-curves that have been suggested in recent years, it should be done not so much by examining the ways in which they have been derived as by seeing what classes of distribution they represent and by noticing carefully the cases of failure and the difficulties of application.

2. As we know from experience that the binomial series actually represents a simple type of probability, it is natural to start from it and treat it, or its limits, as a part of any system; it must, in fact, be a special case of any more general type that may be evolved.

We can proceed either by building up a curve on assumptions which it seems natural to adopt or by taking a more complex series than the binomial (e.g. the hypergeometrical) and in either case an expression might be reached having greater generality than the binomial. But it must be remembered that the ultimate justification of any evolved formula rests mainly on its breadth of application to statistics which may reasonably be described as chance distributions. Such application is an important test of the fundamental assumptions that were adopted in reaching the formula, for it must be admitted that the plausibility of the initial statements would be poor defence of a curve which broke down whenever it was put to a practical test.

The well-known "normal curve of error", with which we dealt on p. 80, was a first step towards finding a simple frequency-curve, but though it works well as a description of the binomial $(p+q)^n$ when p is approximately equal to q or when n is large, it is unsatisfactory in other cases. In actuarial work these cases frequently arise. At the ages attained by the majority of lives assured in any assurance office the rate of mortality or probability of a person dying in a year is small and the frequency distribution giving the number of deaths happening in successive years out of 50 cases, say, when $q = \cdot02$ and $p = \cdot98$, would not be satisfactorily described by the normal curve of error. It is true in a sense that the "normal curve" is a law of great numbers, but if it can only deal with cases resting on such a basis it cannot have a large sphere of action in practical statistics and it can hardly be expected to be of value when a series is more like the hypergeometrical than the binomial.

3. It is this failure of the "normal curve" that has led to the work of Pearson, Thiele, Charlier, Edgeworth, Bruns, Kapteyn and others, and the curves suggested by these writers are of considerable interest to all students of statistical mathematics. In this chapter we shall indicate how far some of these curves fit the statistics that arise in practice; how far, in fact, they graduate the rough figures obtained from the collected facts, and where they break down.

Before proceeding, however, it will be necessary to discuss briefly the suggested types. We may also mention an old difficulty in practical work of this nature, namely, that statistics are seldom obtained from strictly homogeneous material. This fact must be taken as one of the typical elements in practice, and if a series can graduate in spite of a small amount of heterogeneity it is, from some points of view, all the more valuable in much of the work that comes to the hands of an actuary or statistician.

4. We may now turn to an expression which we will call **Type A**, namely

$$F(x) = \phi_0(x) - \frac{1}{3!}\frac{\mu_3}{\sigma^3}\phi_3(x)$$
$$+ \frac{1}{4!}\left(\frac{\mu_4}{\sigma^4} - 3\right)\phi_4(x) - \frac{1}{5!}\left(\frac{\mu_5}{\sigma^5} - \frac{10\mu_3}{\sigma^3}\right)\phi_5(x) + \ldots$$

where $\phi_0(x) = \frac{1}{\sigma\sqrt{(2\pi)}}e^{-x^2/2\sigma^2}$ and $\phi_n(x) = \sigma^n\frac{d^n}{dx^n}\phi_0(x)$

So that, if $\sigma = 1$, i.e. if we measure in terms of the standard deviation,

$$\phi_3(x) = (3x - x^3)\,\phi_0(x)$$
$$\phi_4(x) = (x^4 - 6x^2 + 3)\,\phi_0(x)$$
$$\phi_5(x) = (-x^5 + 10x^3 - 15x)\,\phi_0(x)$$

In applying these expressions x is measured throughout from the mean in terms of the standard deviation: the measures used in *Tables for Statisticians*. It may be mentioned that the coefficients in round brackets in the equation for Type A as set out above are the third, fourth and fifth semi-invariants.

In *Tables for Statisticians* (Pt II, Tables V–VII)

$$\tau_{n+1}(h) = \frac{(-1)^n}{\sqrt{(n+1)!}} \cdot \frac{d^n}{dh^n}\left(\frac{1}{\sqrt{2\pi}}\,e^{-h^2/2}\right)$$

and when using these tables we write F as

$$\frac{N}{\sigma}\{\tau_1(h) + {\cdot}81649658\sqrt{\beta_1}\,.\,\tau_4(h) + {\cdot}45643546(\beta_2 - 3)\tau_5(h) + \ldots\ldots\}$$

This series has been discussed by many writers, especially on the Continent[*], and it may be regarded as the use of the

---

[*] Gram, Thiele, Charlier, Bruns, etc. In a memoir entitled *Researches into the Theory of Probability* (Meddelanden Lunds Astronomiska Observatorium, 1906), C. V. L. Charlier gives several numerical examples and many useful notes. J. P. Gram, on p. 94 of *Om Rækkendriklinger, bestemte ved mindste Kvadraters Methode* (Copenhagen, 1879), says that Oppermann had suggested the formula some time before.

"normal curve" as a generating function. It has, naturally, a greater range of applicability than the "normal curve", but it is not of service in the more extremely skew cases, and it has been suggested by C. V. L. Charlier that, in such circumstances, an expression **Type B** should be used. This is

$$F(x) = B_0 \psi(x) + B_1 \psi_{(x)}^{\mathrm{I}} + B_2 \psi_{(x)}^{\mathrm{II}} + \dots$$

where $\psi(x) = e^{-m} \dfrac{\sin \pi x}{\pi} \left[\dfrac{1}{x} - \dfrac{m}{1!\,(x-1)} + \dfrac{m^2}{2!\,(x-2)} - \dots \right]$

and $\psi_{(x)}^{\mathrm{I}} = \psi(x) - \psi(x-1)$, i.e. $\Delta\psi(x-1)$ and values of $\psi(x)$ for $x < 0$ are assumed to be zero. Similarly $\psi_{(x)}^{N} = \Delta\psi_{(x-1)}^{N-1}$.

In the limit when m is an integer $\psi(x)$ becomes $e^{-m} m^x/x!$. This expression is already well known in the theory of probability as Poisson's series—the "normal curve" is sometimes spoken of as a "law of great numbers" and the Poisson series as a "law of small numbers". Type B uses $e^{-m} m^x/x!$ as a generating function similarly to the way in which Type A uses the "normal curve".

5. The fitting of Type B presents certain special difficulties as alternative methods are available, but we may as a preface to them point out that if we fit $e^{-m} m^x/x!$ by moments using all integral values of x from $x = 0$ to $x = \infty$ we obtain

$$\mu_2 = m \qquad \mu_3 = m \qquad \mu_4 = 3m^2 + m$$

or
$$\beta_1 = \beta_2 - 3 = 1/m$$

This, however, assumes a system of ordinates, unit distance apart, and we know that in practical statistical work these assumptions limit us unduly.

We can, however, write

$$F(xw + c) = B_0 \psi(x) + B_1 \psi_{(x)}^{\mathrm{I}} + B_2 \psi_{(x)}^{\mathrm{II}} + \dots$$

which implies that owing to w we have generalized the unit of grouping and owing to c the point from which x is reckoned is also generalised.

In this form Charlier suggests four methods of fitting and remarks that the series usually becomes more convergent if we arrange constants so that $B_1 = B_2 = B_3 = 0$.

(131)

(1) Assume $w = 1$ and $c = 0$, that is, revert to the original form and choose m so that B_1 vanishes, and since $B_0 = N$ we can reach

$$2!\, B_2 = N(\mu_2 - b)$$
$$3!\, B_3 = N(-\mu_3 + 3\mu_2 - 2b)$$
$$4!\, B_4 = N(\mu_4 - 6\mu_3 - 6b\mu_2 + 11\mu_2 + 3b^2 - 6b)$$

where b is the distance from the origin to the mean. This method can be used when we can anticipate that m will not differ greatly from b.

(2) Assume $w = 1$ and calculate c as an unknown constant, choosing it and m so that B_1 and B_2 vanish.

$$c = b - \mu_2 \qquad 3!\, B_3 = N(\mu_2 - \mu_3)$$
$$m = \mu_2 \qquad 4!\, B_4 = N(\mu_4 - 3\mu_2^2 - 6\mu_3 + 5\mu_2)$$

(3) Find m, w and c so that $B_1 = B_2 = B_3 = 0$.

$$w = \mu_3/\mu_2 \qquad B_0 = N/w$$
$$m = \mu_2^3/\mu_3^2 \qquad B_4 = \frac{N}{24w^5}\left(\mu_4 - 3\mu_2^2 - \frac{\mu_3^2}{\mu_2}\right)$$
$$c = b - \mu_2^2/\mu_3$$

This method usually gives w very small values and m very large values when μ_3 vanishes, so it is only applicable in markedly skew cases.

(4) Fix c arbitrarily and find m and w so that $B_1 = B_2 = 0$.

$$m = (b - c)^2/\mu_2$$
$$w = \mu_2/(b - c)$$
$$B_0 = N/w$$
$$w^3 3!\, B_3 = B_0(w\mu_2 - \mu_3)$$
$$w^4 4!\, B_4 = B_0(\mu_4 - 3\mu_2^2 + 5w^2\mu_2 - 6w\mu_3)$$

It seems unnecessary to give the work in detail leading up to the various sets of equations. Tables of $e^{-m} m^x/x!$ will be found in *Tables for Statisticians*.

6. F. Y. Edgeworth* has used a series similar to Type A, namely:

$$e^{-\frac{k_1}{3!}\left(\frac{d}{dx}\right)^3+\frac{k_2}{4!}\left(\frac{d}{dx}\right)^4-\cdots}\phi_0(x)$$

where k_1, k_2, etc. are the third, fourth, etc. semi-invariants. Expanding the exponential, we reach

$$\frac{N}{\sigma}\{\tau_1(h) + \cdot 81649658\sqrt{\beta_1}\tau_4(h) + \cdot 98601330\beta_1\tau_7(h) + \ldots$$
$$+ \cdot 45643546(\beta_2 - 3)\tau_5(h) + \ldots + \text{etc.}\}$$

Arithmetically the difference between this series and Type A is usually small. Type A does not include the τ_7 term which arises from $k_1^2\frac{1}{2!}\frac{1}{3!}\frac{1}{3!}\frac{d^6}{dx^6}$. Later terms would also differ, but the expansion shown assumes that we shall not use more than four moments and that τ_9 etc. terms can be ignored.

7. It is possible to use other expressions, e.g. Type III, instead of a normal curve as a generating function. A necessary condition for a frequency function is that it must not produce negative frequencies and the reader who wishes to pursue this part of the subject may be referred to a lecture by Professor Steffensen giving an interesting account from first principles.†
For a general discussion of Edgeworth's and the A series and the theory underlying them the reader should study the papers to which reference has already been made and also Professor H. Cramèr's paper "On the composition of elementary errors" in *Skandinavisk Aktuarietidskrift*, 1928, p. 13 etc. and p. 141 etc.

It is not, however, pretended that the curves and series set out above exhaust the suggestions that have been made, but they may be taken to represent the methods that have

* Edgeworth contended that his equation was unique in its character and theoretical basis. It avoids the negative frequencies which may arise with Type A and are unjustifiable in theory. This last point will be brought out in the numerical examples. *Trans. Camb. Phil. Soc.* 1905 (Law of Error), *J. Roy. Statist. Soc.* 1906 (Generalised Law of Error).

† J. F. Steffensen, *Some recent researches in the Theory of Statistics and Actuaria Science* (Cambridge University Press, 1930, Third Lecture).

received most general support, and the examples we shall give do not go beyond them. We may, however, mention that it has been suggested that graduations should be made by writing $y = e^{-\frac{1}{2}[f(x)]^2}$. This way of using the "normal curve" has been called the "Method of Translation" and in its most general form is arbitrary. In practice the form of $f(x)$ must be restricted and certain special cases have been studied* but the method seems to be open both to practical and theoretical objections, and it will not be discussed in detail.

8. Numerical Examples.

Example I

(Symmetrical curve not capable of satisfactory graduation by the normal curve of error.)

Observations	Pearson's Type II	Type A	Edgeworth	Normal curve
11	14	15	16	20
116	109	106	106	95
274	286	284	285	270
451	433	437	436	456
432	433	437	436	456
267	285	283	284	270
116	109	106	106	95
16	14	15	16	20

In this case all the curves except the normal give excellent graduations. We have not used Type B because Charlier apparently only adopts it when Type A is unsuccessful. He does not give a statistical criterion to show when A or B should be used and it is difficult to see how such a criterion can be evolved. The solution of his Type A does not lead to imaginary quantities when Type B should have been used, in the way that Pearson's Type I, for example, does when it is inapplicable. In reaching Type A and the Edgeworth graduation we have used the terms involving A_4 and k_2 respectively. A_4 is used here for the coefficient of $\phi_4(x)$. Similarly hereafter with A_n. Notice that A_n involves μ_n but may also involve other μ's.

* See Edgeworth, *J. Roy. Statist. Soc.* vol. LXI; Kapteyn, *Skew Frequency Curves in Biology and Statistics* (Groningen, 1903); or Bowley, F. Y. *Edgeworth's Contributions to Mathematical Statistics*, 1928.

Example II

(A distribution which is not markedly skew)

Observations	Pearson's Type III	Type A	Edgeworth
3	4	5	4
20	17	22	17
38	42	47	42
63	59	60	59
51	53	50	53
29	33	27	32
21	15	13	15
4	5	4	6
0	1·4	1	2
1	0·4	...	1

In each case three moments have been used. The observations and Edgeworth's graduation are taken from Edgeworth's paper, "The generalised law of error". Type A is the least successful.

Example III

(A distinctly skew distribution)

Observations	Pearson's Type I	Type A	Type B	Edgeworth
...	...	-2	...	...
...	...	1	...	1
...	...	8	...	9
...	2	25	12	30
64	67	53	64	64
116	116	90	104	102
140	138	125	129	130
145	139	145	134	135
134	128	143	128	130
106	110	123	116	111
82	89	93	93	92
72	69	65	73	73
49	51	44	53	53
37	35	31	36	36
25	24	23	25	20
13	15	16	14	10
10	9	10	10	4
5	5	5	5	...
2	2	2	2	...
0·4	1	1	1	...

Pearson's figures come from his *Chances of Death** and

* *Chances of Death*, i, 74 (London, 1897).

(135)

Edgeworth's from his "Generalised law of error". Each of these graduations was obtained with four moments. Clearly Pearson's Type I is the best and Type B the next best graduation. We do not think Charlier would use Type A in such a case. In fitting his Type B there are, however, many difficulties owing to the fact that he gives us four approximate methods of application; this is an objection which may be surmounted in the future, but makes Type B awkward at present. The other points to be noticed in these graduations are the negative frequency in Type A and the 40 cases in Edgeworth's graduation which have no case corresponding to them in the data. Edgeworth, however, has remarked that he only aims at the main body of the curve and does not much concern himself with the tails, but one cannot help feeling that the main body must be understated if one tail possesses an excess of 40 out of 1,000 cases and the other tail is in defect by only 20.

Example IV

(J-shaped curve)

Observations	Pearson	Type B
133	136·9	134·9
55	48·5	51·6
23	22·6	22·5
7	9·6	9·5
2	3·4	2·9
2	·8	·6

The Type B curve is given by Charlier in *Researches into the Theory of Probability*. The Type B curve gives a slightly better graduation, but the agreement is close in both cases. The example is not conclusive as to J-shaped curves, but shows that Type B can graduate them successfully. The particular example has only six groups, and with a curve of something like the right shape and three constants we are likely to reach close agreement. Edgeworth's curve is unsuitable. A graduation by Type A has been given elsewhere, but though it apparently graduates the figures the curve is not J-shaped.

Example V

(Series which is nearly symmetrical)

Observations	Pearson's Type IV	Type A	Edgeworth
10	6	4	3
13	16	14	10
41	49	46	34
115	135	126	110
326	321	306	298
675	653	637	662
1,113	1,108	1,108	1,164
1,528	1,535	1,563	1,603
1,692	1,712	1,753	1,747
1,530	1,522	1,548	1,510
1,122	1,074	1,075	1,024
611	604	589	571
255	274	256	263
86	102	92	104
26	32	29	37
8	8	7	12
2	2	2	2
1	1	1	1
1	...	...	...

These graduations give similar results and need no comment.

Example VI

(Distribution having two maxima)

Data	Pearson's Type II	Type A	Edgeworth
10	3	26	4
78	96	74	34
193	191	156	135
286	261	262	270
303	304	354	363
291	319	390	390
303	304	354	363
286	261	262	270
193	191	156	135
78	96	74	34
10	3	26	4

This is an imaginary example giving a double-humped distribution. It was formed from Type A by putting $A_3 = 0$ and $A_4 = \cdot 09$; the series being

$$-4, \ -19, \ -53, -76, \ +103, \ +783, \ +1929, \ +2855, \ \text{etc.}$$

Negative frequencies, which are meaningless, were discarded

and the data cut down and graduated. The interesting feature is that Type A from which the data were formed gives a poor agreement. This is due to the negative frequencies and the integration for moments from $-\infty$ to $+\infty$. Negative frequencies are somewhat objectionable in themselves; they are still more objectionable when they influence curve fitting to the large extent shown in this example.

Example VII

We have remarked that there is a difficulty in choosing a solution to Type B, but its graduating power compared with other formulae can be indicated by setting out a few examples of the forms taken by $e^{-m}m^x/x!$ from *Tables for Statisticians*.

For comparison I have added examples of Pearson's Type III, though it must not be supposed that either set is meant to give the closest agreement with the other that it would be possible to make; they have merely been taken to give an idea of the range of application. By bringing in terms involving $\psi_{(x)}^N$ we can increase the range of Type B and by using the whole of Pearson's system we cover a wider range than that of his Type III.

TYPE B			PEARSON'S TYPE III		
I	II	III	I	II	III
368	111	45	387	63	31
368	244	140	386	279	149
184	268	217	160	285	230
61	197	224	47	189	218
15	108	173	15	102	160
3	48	107	4	49	101
1	18	55	1	21	56
...	6	25	...	9	29
...	2	10	...	3	14
...	...	3	...	...	6
...	...	1	...	...	3
...	...	...	...	...	1

9. The few examples we have given will be of help in bringing out the comparison of the types of curves with which we have been dealing.

The Pearson-type curves will graduate satisfactorily all the examples we have taken, but cannot reproduce the double hump of our imaginary data (Example VI). They will graduate symmetrical, slightly skew and very skew distributions and also J and U-shaped distributions. They have been fitted in various circumstances and are satisfactory from the point of view of agreement. The arithmetic involved is, however, very heavy, but the curves are the most useful of those now considered.

Type A gives numerically the least work, but it does not graduate satisfactorily very skew or J and U-shaped distributions and it has therefore a smaller vogue. If, however, it is combined with Type B as Charlier suggests, J-shaped and skew distributions can be graduated. We have found some difficulty in applying Type B, for Charlier does not give much help in deciding which of his four methods of fitting should be followed in a particular case, and we feel that the graduation capacity of this type may be greater than our trials with it justify us in thinking at present. It would clearly be impossible to improve on its graduation in Example IV, but Example III and two examples given by Charlier in his *Researches into the Theory of Probability* are less fortunate.

Edgeworth's curve can, roughly speaking, graduate the same distributions as Type A.

10. We may now refer to two difficulties in connection with Edgeworth's curve and with Type A respectively which have already been mentioned. In Example III we found that 40 out of 1,000 in Edgeworth's graduation have no observations corresponding to them and we remarked that it seemed a large excess; the reproduction of the exact number of observations is not only a practical necessity, but is assumed by the method of moments. If, therefore, a large number of cases falls outside the observations, we must either say that the total frequency is not reproduced or that the frequencies are misplaced; in either case the main body must be artificially reduced below the amount shown in the original data. In slightly skew distributions the frequencies are satisfactorily reproduced and many of the graduations of such material are excellent, but the method can hardly be considered satisfactory as a general formula until some method of overcoming the difficulty mentioned above has been found.

The difficulty in connection with Type A is the large part that negative frequencies play in some of the less symmetrical graduations. If a negative frequency occurs, have the positive frequencies been overstated? The defence of such negatives is that further terms of the series would put things right, but it is hard to see the justification for basing much argument on constants derived from the higher moments which are liable to large variations and are unreliable. It is also unsatisfactory that a curve cannot reproduce itself even approximately, and the result of our Example VI is disappointing; probably however it would be well to consider such cases as relating to heterogeneous material and therefore more suitable for representation by two or more superimposed curves.* If Type A or Edgeworth's curve and their moments could be integrated from $-a$ to b instead of from $-\infty$ to ∞, the difficulties could be overcome to some extent; but, failing that, it would seem necessary to limit the range of applicability to the less abnormal distributions. An approximate method of fitting from $-a$ to ∞ has been given†, but the results are not quite so good as Pearson's Type III.

11. If the reader makes any extensive trial with series for the purpose of graduation, he will find occasionally that the coefficients of successive terms are such as to imply that the series may not be convergent. This is closely connected with the difficulty mentioned in the preceding paragraph.

---

* We are doubtful if it is statistically possible ever to produce a double hump with Type A or Edgeworth's curve if the ordinary $-\infty$ to ∞ integration is performed, because the relative values of the second and fourth moments required by the coefficient in the formula would seem impossible.

† E. C. Rhodes, *J. Roy. Statist. Soc.* 1925, pp. 576 et seq.

CHAPTER VII

CORRELATION

1. We say that tall men have longer legs than short men, that the older a bachelor the less likely he is to marry and have children, that a man marrying late in life usually takes a wife who is older than the wife of a man marrying early, or, to take an example from life assurance practice, that, when endowment assurances are grouped according to the unexpired term, the mean ages at maturity increase with the unexpired term. All these statements express in different words the fact that there is some causal relationship, or correlation, between the height of a man and the length of his legs, between the ages of husband and wife or between the age at maturity of endowment assurances and the unexpired term. The statements are, however, in general terms; they do not help us to decide whether one relationship is closer than another; they do not supply any scale of correlation. The object, in statistical work, is to find a measure; we have a scale for measuring probability and similarly we want a scale for measuring correlation.

This suggests that if there is no correlation our scale ought to measure zero and, just as certainty is indicated by a probability of unity, so we may call our correlation unity when the relationship is as close as possible. There is, however, one point where the analogy between probability and correlation breaks down; there is no such thing as negative probability, but we can easily see that we can have negative correlation, for we may have two things, A and B, which increase together like the ages of husband and wife, or two things, C and D, one of which increases as the other decreases like the age of a bachelor and the number of children born from subsequent marriages.

2. With this introduction we may set down a definition of correlation in the following words: "two measurable characteristics, *A* and *B*, are said to be correlated when, with different values x of A, we do not find the same value y of B equally likely to be associated." In other words, certain values of B are more likely to occur with the value x than others. If they were not, correlation would be absent, or, to take a specific case, if men marrying at 20, or at 30, or at 60, or at any other age always married women of 40, there would be no correlation. On the other hand, the correlation would be perfect if every man had to marry a woman exactly n years his junior.

Unexpired term of endowment assurances (centre of group of 5 terms)	Central age at maturity										Total	Mean maturity age for the row
	30	35	40	45	50	55	60	65	70	75		
2	2	...	...	2	26	6	14	6	...	...	56	53·75
	24	20	16	12	8	4	0	4				
7	1	1	2	6	62	36	40	22	2	...	172	55·03
	18	15	12	9	6	3	0	3	6			
12	...	2	9	17	117	99	127	52	8	1	432	55·85
		10	8	6	4	2	0	2	4	6		
17	3	...	6	24	145	155	237	84	11	...	665	56·59
	6	5	4	3	2	1	0	1	2	3		
22	...	1	...	3	133	167	271	78	20	1	674	57·58
	0	0	0	0	0	0	0	0	0	0		
27	...	...	...	9	90	123	231	71	11	3	538	57·88
				3	2	1	0	1	2	3		
32	...	...	...	1	11	49	127	49	8	2	247	59·94
				6	4	2	0	2	4	6		
37	...	...	...	...	...	6	49	22	...	...	77	61·04
						3	0	3				
42	...	...	...	...	...	2	2	3	...	1	8	62·50
						4	0	4	8	12		
47	...	...	...	...	...	...	...	1	...	...	1	65·00
							0	5				
Total	6	4	17	62	584	643	1098	388	60	8	2,870	
Mean unexpired term for column	10·3	13·2	13·2	16·1	17·2	20·1	21·9	21·7	21·5	27·6		

NOTES. For explanation of small numbers, see § 10.

A column or row is called an array. The middle value of the variable with which the row is associated is called its type, so that the third column (i.e. that headed 40) would be called the y-array of type 40, and the fourth row would be called the x-array of type 17. The word 'type' is sometimes omitted.

3. The statistical aspect of the problem is exemplified in the above table of double entry which gives particulars of 2,870 endowment assurances grouped according to unexpired term.

A little examination of the table shows that correlation is present, for we notice that the figures in the column giving the mean maturity age for each row increase steadily from 53·75 to 65, while the term increases from 2 to 47. Similarly, the mean unexpired terms increase from 10·3 to 27·6 as the age at maturity increases from 30 to 75. The two sets of figures are indicated in the diagram, p. 144. Now let us imagine that there was no correlation, then the means of the columns would have been independent of the other function, that is, we should have found the same mean for each column. When plotted on a diagram the means would have run horizontally. This suggests that, perhaps, correlation might be measured by the slope of a straight line drawn through the means, and we may follow up this idea by fitting a straight line ($y = a_2 + b_2 x$) to the correlation table and seeing what we can gather from the result.

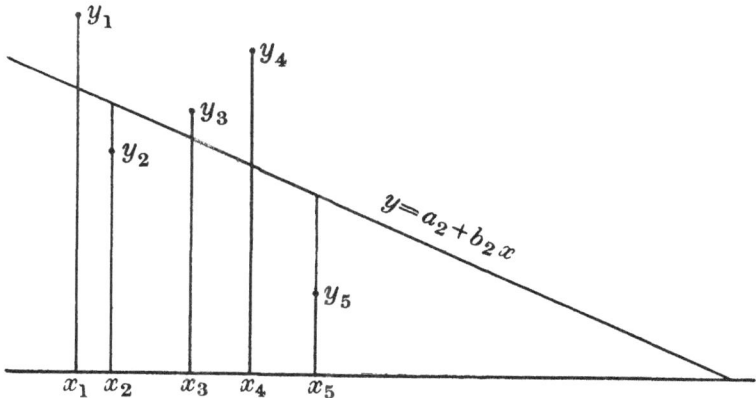

4. When we were fitting curves to frequency distributions we used the Method of Moments, and the following proof adopts the same principle.

(143)

Let $x_1 y_1$, $x_2 y_2$, etc. be associated deviations, and let

$$y = a_2 + b_2 x$$

be the straight line used in the graduation; then the graduated figure corresponding to x_1 is $a_2 + b_2 x_1$.

Now, if we proceed as we did in fitting frequency-curves by

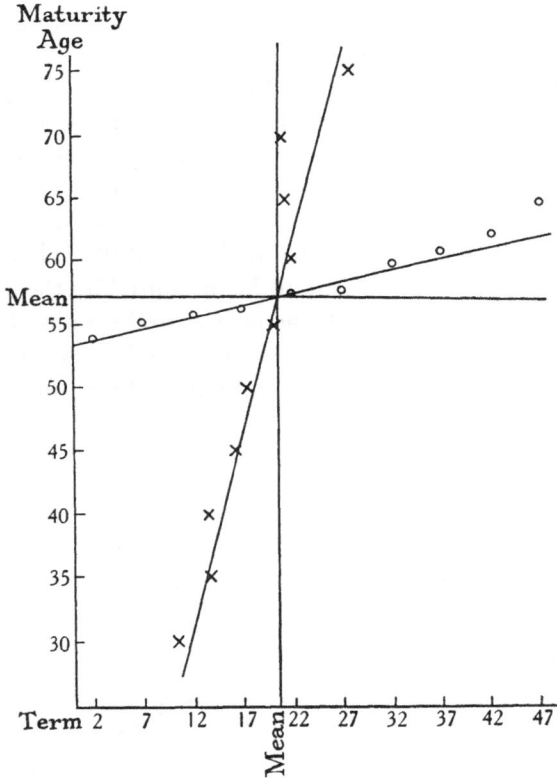

NOTE. The mean unexpired terms corresponding to actual central ages at maturity are shown × and the mean central ages at maturity corresponding to actual unexpired terms are shown o.

The diagram is arranged so that the standard deviation of the maturity ages is represented by the same length as the standard deviation of the unexpired terms and, consequently, the angles formed by the two regression lines with their respective axes are equal. The tangent of this angle in each case is $r(\cdot 254)$.

the method of moments, we make the graduated and un-graduated areas, means, etc. equal, or

$$(a_2 + b_2 x_1) + (a_2 + b_2 x_2) + \ldots = y_1 + y_2 + \ldots$$

or $$Na_2 + b_2 S'(x) = S'(y)$$

And $$(a_2 + b_2 x_1) x_1 + (a_2 + b_2 x_2) x_2 + \ldots = x_1 y_1 + x_2 y_2 + \ldots$$

or $$a_2 S'(x) + b_2 S'(x^2) = S'(xy)$$

where $S'(x)$, being the sum of all the x's, gives the first moment of the x's, $S'(y)$ the first moment of the y's, $S'(x^2)$ the second moment for the x's, and $S'(xy)$ a moment in which any frequency is multiplied by the product of the distances in the x and y directions.*

If these moments are now transferred to the mean, as was done in fitting the frequency-curves, we have

$$Na_2 = 0 \quad \text{or} \quad a_2 = 0$$

and $$b_2 S(x^2) = S(xy) \quad \text{or} \quad b_2 = \frac{S(xy)}{S(x^2)}$$

But we have already seen that the second moment of the whole frequency (N) is $N\sigma_1^2$; therefore

$$b_2 = \frac{S(xy)}{N\sigma_1^2}$$

and $$\bar{y} = \frac{S(xy)}{N\sigma_1^2} x$$

If we now write $S(xy) = N\sigma_1\sigma_2 r$, we have

$$\left. \begin{array}{l} \bar{y} = r\dfrac{\sigma_2}{\sigma_1} x \\[2ex] \bar{x} = r\dfrac{\sigma_1}{\sigma_2} y \end{array} \right\}$$

where r will represent the statistical measure of correlation (coefficient of correlation) between the x's and y's and the second equation has been evolved similarly to the first.

* Cp. Table II, p. 16. The frequency 29 is multiplied by the appropriate value (-4). It would be the same thing if we took the distance (-4) of each of the 29 cases and added these twenty-nine (-4)'s together.

5. At first sight it may appear that the two equations just given, showing the relationship between x and y, are not consistent. It must, however, be remembered that the first, $\bar{y} = r\dfrac{\sigma_2}{\sigma_1}x$, gives the mean values of y corresponding to particular values of x (indicated by the insertion of a bar over the y), while the second gives the mean values of x corresponding to particular values of y. To take a simple case as an example, assume that $\sigma_1 = \sigma_2 = 1$ and that $r = \cdot 1$, then if $x = 0$ the mean of the y's corresponding to this value of x is 0, and if $x = 20$ the mean of the y's will be 2. When we turn the matter round, however, we cannot, of course, assert that the mean of the x's corresponding to $y = 2$ is 20; it will be $\cdot 2$.

6. After this preliminary remark we may return to the two equations and consider how it is that r is a measure of correlation and whether it can always be treated as a satisfactory measure. We can best see that r is a measure of correlation by rewriting the equation $\bar{y} = r\dfrac{\sigma_2}{\sigma_1}x$ in the form $\dfrac{\bar{y}}{\sigma_2} = r\dfrac{x}{\sigma_1}$ or $\bar{Y} = Xr$, and we can then interpret it as giving one characteristic in terms of the other where the mean is the origin (this is due to referring moments to the mean in the proof) and the unit of measurement is the standard deviation in each case. In this form we see at once that as one characteristic (X) increases the mean $(\bar{Y})$ of the corresponding series of the other characteristic increases to an extent which depends on the value of r; while if r is negative $\bar{Y}$ decreases. It is only if r is unity that the increments of X and $\bar{Y}$ become equal and absolute correlation is reached. If $\bar{Y}$ remains constant as the value of X increases, the definition at the beginning of this chapter tells us that there is no correlation, and r in this case is zero as can easily be seen from the equation $\bar{Y} = Xr$. We have anticipated that our scale for measuring correlation should run from -1 to $+1$, but we may accentuate the fact that a large negative value does not mean that the two characteristics do not vary together but only that increases in the one correspond with decreases in the other; the numerical value of r indicates the extent to which variations in the two characteristics correspond. This indication is satisfactory provided the means, when plotted in a diagram such as that on p. 144, fall approximately in a straight line (i.e. "regression"* is linear). Distinct deviations from linearity are not so common as might be supposed, but if they are very marked in any case, r ceases to be a satisfactory measure of the correlation.

7. We may take this opportunity of removing another difficulty that is sometimes met. Some students have a doubt which is best shown by the question: " How can there be perfect correlation when one thing is always smaller than another? " As an example we may take the correlation between the lengths of a man's right arm and his left arm; here the coefficient of correlation would be practically unity, and since each characteristic is measured from its own mean, and in terms of its own standard deviation, the coefficient would not be decreased if every left arm was a certain number of inches shorter than the right or if it bore a fixed relation in length, say 99/100, to the right arm.

8. It is now necessary to discuss the arithmetical calculations and if we look back at the formulae at the end of § 4 we see that we require two standard deviations and a value for $S(xy)$. We have already seen how standard deviations are obtained and it will be remembered that when the calculation of moments was discussed we found that, though they were required about the mean, it was best in practice to take them about some point fixed arbitrarily so as to avoid fractions and then adjust the results afterwards. The values of the σ_1 and σ_2 can, of course, be found with the help of the formula on p. 57, viz. $\nu_2 = \nu_2' - d^2$. The deduction of $\frac{1}{12}$ from the second moment should be made for the same reason and in the same cases as in frequency-curve fitting.

---

* The term "regression" was adopted by Francis Galton in connection with the study of heredity; it indicates the way the children of particular parents tend to "step back" to the ordinary population mean.

With regard to the product moment we have

$$S(x'y') = S(x+d_1)(y+d_2)$$
$$= S(xy) + d_1 S(y) + d_2 S(x) + N d_1 d_2;$$

or since $\quad S(x) = S(y) = 0$

$$S(xy) = S(x'y') - N d_1 d_2$$

where $S(x'y')$ is calculated about a point distant d_1 from the mean of the x's and d_2 from the mean of the y's.

9. The statistical example on p. 142 can now be worked through. It will be found to make the proofs and methods given above much easier to grasp.

A point about which moments are to be calculated is first fixed, say the middle of the group corresponding to maturity age 60 and unexpired term 22 years, and for the present the calculations are made about this point. The following table shows the calculation of the mean and the second moment of the totals of the y-arrays, i.e. the totals at the bottom of the table, because columns are y-arrays and rows x-arrays:

Frequency	x'	Frequency $\times x'$	Frequency $\times (x')^2$
6	-6	36	216
4	-5	20	100
17	-4	68	272
62	-3	186	558
584	-2	1,168	2,336
643	-1	643	643
1,098	0	$-2,121$	...
388	1	388	388
60	2	120	240
8	3	24	72
$2,870 = N$		$+ 532$	4,825
		$-1,589$	

$$d_1 = -\frac{1589}{2870} = -\cdot55366$$

Hence, the mean age $= 60 - 2\cdot7683 = 57\cdot2317$, because the unit of grouping is 5 years.

$$\sigma_1^2 = \frac{4825}{2870} - d_1^2 - \tfrac{1}{12} \text{ (Sheppard's adjustment)}$$

$$= 1{\cdot}37465 - {\cdot}083$$

$$= 1{\cdot}29132$$

$$\therefore \quad \sigma_1 = 1{\cdot}13637$$

Treating the rows in the same way, the following table was formed:

Frequency	y'	Frequency $\times y'$	Frequency $\times (y')^2$
56	-4	224	896
172	-3	516	1,548
432	-2	864	1,728
665	-1	665	665
674	0	-2,269	...
538	1	538	538
247	2	494	988
77	3	231	693
8	4	32	128
1	5	5	25
2,870 = N		+1,300	7,209
		- 969	

$$d_2 = -\frac{969}{2870} = -{\cdot}33763$$

$\therefore$ Mean unexpired term $= 22 - 1{\cdot}68815 = 20{\cdot}31185$

$$\sigma_2^2 = \frac{7209}{2870} - d_2^2 - \tfrac{1}{12}$$

$$= 2{\cdot}31453$$

and $\qquad \sigma_2 = 1{\cdot}52135$

10. The value of $S(xy)$ is formed with the help of the numbers in very small type appearing under the frequencies in the correlation table. The frequency 62 in the 50 column, for instance, is distanced three spaces upwards and two sideways from the arbitrary origin, so the value of $x' y'$ by which it has to be multiplied is $3 \times 2 = 6$, as shown in the small type. The other figures are obtained in like manner, but the sign must be borne in mind. Any value from the left-hand upper division of the table, or in the lower right-hand division, will be positive, because the frequency will be multiplied by a product of an x and y having like signs; while any value from the other divisions will be negative, because the x and y by which the frequencies are multiplied are of opposite signs.

The calculation of the product moment is as follows:

Frequencies	$x'\,y'$	Total of frequencies (f)	$f \times x'\,y'$
$155 + 71 - 84 - 123$	1	$+19$	$+19$
$145 + 99 + 11 + 49 - 11 - 52 - 49 - 90$.	2	102	204
$24 + 36 + 3 + 22 - 22 - 6 - 9$ . .	3	48	144
$6 + 6 + 8 + 3 - 6 - 8 - 11 - 2 + 117$.	4	113	452
1	5	1	5
$3 + 17 + 62 + 2 - 1 - 1 - 2$ . . .	6	80	480
$9 + 26$	8	35	280
6	9	6	54
2	10	2	20
$2 + 2 + 1$	12	5	60
1	15	1	15
1	18	1	18
2	24	2	48
			1,799

$$S(xy) = S(x'y') - Nd_1 d_2$$
$$= 1799 - Nd_1 d_2$$
$$= 1262 \cdot 51$$
$$r = \frac{S(xy)}{N\sigma_1\sigma_2} = \frac{1262 \cdot 51}{2870 \times 1 \cdot 13637 \times 1 \cdot 52135}$$
$$= \cdot 254.$$

The coefficient of correlation between age at maturity and the unexpired term of endowment assurances is $\cdot 254$.

The equation representing the one function in terms of the other is
$$\bar{x} = r\frac{\sigma_1}{\sigma_2}y$$
$$= \cdot 190y$$

where all measurements are made from the mean and the unit is 5 years. The line drawn in the figure gives this result.

11. An alternative method similar to the summation method given in §9, Chapter III for moments can be conveniently used in connection with correlation tables. Taking the same example, we obtain from the given table another in the same form, giving the y sum of it by summing each column continuously, and then form a third table by summing the second table across continuously.

Table of the y-sum of Correlation Table

Unexpired term of endowment assurances	Central age at maturity										
	30	35	40	45	50	55	60	65	70	75	Totals
2	6	4	17	62	584	643	1,098	388	60	8	2,870
7	4	4	17	60	558	637	1,084	382	60	8	2,814
12	3	3	15	54	496	601	1,044	360	58	8	2,642
17	3	1	6	37	379	502	917	308	50	7	2,210
22	0	1	0	13	234	347	680	224	39	7	1,545
27	0	0	0	10	101	180	409	146	19	6	871
32	0	0	0	1	11	57	178	75	8	3	333
37	0	0	0	0	0	8	51	26	0	1	86
42	0	0	0	0	0	2	2	4	0	1	9
47	0	0	0	0	0	0	0	1	0	0	1
Totals	16	13	55	237	2,363	2,977	5,463	1,914	294	49	13,381

Table of x-sum of above Table, i.e. Table giving all cases for xy group and over in Correlation Table

Unexpired term of endowment assurances	Central age at maturity										
	30	35	40	45	50	55	60	65	70	75	Totals
2	2,870	2,864	2,860	2,843	2,781	2,197	1,554	456	68	8	18,501
7	2,814	2,810	2,806	2,789	2,729	2,171	1,534	450	68	8	18,179
12	2,642	2,639	2,636	2,621	2,567	2,071	1,470	426	66	8	17,146
17	2,210	2,207	2,206	2,200	2,163	1,784	1,282	365	57	7	14,481
22	1,545	1,545	1,544	1,544	1,531	1,297	950	270	46	7	10,279
27	871	871	871	871	861	760	580	171	25	6	5,887
32	333	333	333	333	332	321	264	86	11	3	2,349
37	86	86	86	86	86	86	78	27	1	1	623
42	9	9	9	9	9	9	7	5	1	1	68
47	1	1	1	1	1	1	1	1	0	0	8
Totals	13,381	13,365	13,352	13,297	13,060	10,697	7,720	2,257	343	49	87,521

The totals in the right-hand column of the upper table give the first sum of the total in the right-hand column of the correlation table, and are the same as the column $x = 30$ in the lower table. The total of the y sum, or of the first column in the xy table, gives the mean of the y's (13,381/2,870), and similarly the sum of the first row gives the mean of the x's

$$(18,501/2,870).$$

The total of the last table gives the xy moment (87,521), and the x standard deviation is found by forming from the first row the series 18501, 15631, 12767, 9907, 7064, 4283, 2086, 532, 76, 8, and summing it, i.e. 70,855. The second moment about the mean can then be found, the numerical work being as follows:

$$x \text{ mean} = \frac{18501}{2870} = 6\cdot4463$$

$$\nu_2 = 2S_3 - d(1 + d)$$

$$= \frac{2 \times 70855}{2870} - 6\cdot4463 \times 7\cdot4463$$

$$= 1\cdot3747$$

Similarly with the y moments

$$y \text{ mean} = \frac{13381}{2870} = 4\cdot6624$$

$$\nu_2 = \frac{2\,(13381 + 10511 + 7697 + 5055 + 2845 + 1300 + 429 + 96 + 10 + 1)}{2870} - 4\cdot6624 \times 5\cdot6624$$

$$= 2\cdot2312$$

The xy moment $= \dfrac{87521}{2870} - 6\cdot4463 \times 4\cdot6624$

$$= \cdot4399$$

Remembering that $\nu_2 - \frac{1}{12}$ (Sheppard's adjustment) $= \sigma^2$ and that the means are, in the above work, measured from the centre of the group $x = 25$, $y = -3$ years, the values just given will be found to agree with those previously obtained by the direct method. The xy moment ($\cdot 4399$) is the same as $\dfrac{1262 \cdot 5}{2870}$, i.e. $S(xy)/N$.

12. We have already remarked (§ 6) that the method we have used for measuring correlation assumes that the means of the rows and of the columns, respectively, lie on straight lines and consequently we must examine a table to see whether this holds. One advantage of the method given in the previous paragraph is that it enables us to get the means of each column and of each row very easily. Remembering:

(1) that the interval between the groups of unexpired terms is 5 years

(2) that the interval between central ages at maturity is 5 years

(3) that the arbitrary origin is the point representing central age at maturity $= 25$ and unexpired term $= -3$

we can get the means of the columns by taking each total in the y-sum table, multiplying by 5, dividing by the number of cases and subtracting 3; thus, for the column with central age 50

$$\frac{2363 \times 5}{584} - 3 = 17 \cdot 23$$

The means of the rows come from the differences of the totals on the right of the next table and thus for unexpired term 2 we have

$$(18501 - 18179) \times 5/56 + 25 = 53 \cdot 75$$

and for unexpired term 17

$$(14481 - 10279) \times 5/665 + 25 = 56 \cdot 59$$

13. There is yet a third way of doing the arithmetical work to reach the coefficient of correlation and, as it is short and relies on one of the series of means, it has a good deal to commend it. The calculation is as follows:

First moment of column (for total frequency in column) about arbitrary origin of columns (i.e. unexpired term 22) (1)	Distance of column from the arbitrary origin of rows (i.e. age 60) (2)	(1) × (2) (3)
− 14	− 6	+ 84
− 7	− 5	+ 35
− 30	− 4	+ 120
− 73	− 3	+ 219
−557	− 2	+ 1,114
− 238	− 1	+ 238
...	0	...
− 26	+ 1	− 26
− 6	+ 2	− 12
+ 9	+ 3	+ 27
		1,799

$$r = \left(\frac{\text{Total of (3)}}{N} - d_1 d_2\right) \Big/ (\sigma_1 \sigma_2)$$

$$= \cdot 254 \text{ as in } \S 10.$$

The unit throughout is 5 years and the easiest way to do the calculation for col. (1) is as shown in the table on p. 155.

There is no need to insert a column for age 60, or a row for term 22, as these are multiplied by zero; they are sometimes worked out for completeness and because they make it easier to apply arithmetical checks which the reader can evolve for himself.

If the reader considers any item in this scheme, e.g. 18 in column headed 40, he will see that it represents 9 cases in the table (p. 142) multiplied by − 2, and when it is, amongst other numbers, taken to col. (1) of the table above, it will be multiplied by − 4; that is, we shall have multiplied 9 by (− 2) × (− 4), i.e. by 8, which is the little figure written under 9 in the table on p. 142.

Before dealing with other examples and methods, it may be well to point out a use to which the particular example might be put. The result in the equation form gives the average age corresponding to each unexpired term. Now, we might weight

(Frequency in column) × (distance from arbitrary origin)

Maturity age	30	35	40	45	50	55	65	70	75
Distance from arbitrary origin	"Minus" products								
− 4	8	...	...	8	104	24	24	...	...
− 3	3	3	6	18	186	108	66	6	...
− 2	...	4	18	34	234	198	104	16	2
− 1	3	...	6	24	145	155	84	11	...
Total minus	14	7	30	84	669	485	278	33	2
	"Plus" products								
+ 1	...	...	...	9	90	123	71	11	3
+ 2	...	...	...	2	22	98	98	16	4
+ 3	...	...	...	...	...	18	66	...	...
+ 4	...	...	...	...	...	8	12	...	4
+ 5	...	...	...	...	...	...	5	...	...
Total plus	...	...	...	11	112	247	252	27	11
Figs. for col. (1)	− 14	− 7	− 30	− 73	− 557	− 238	− 26	− 6	+ 9

each entry with Lidstone's Z's,* or with the temporary annuities, then work out an equation in each case, and get new series of average ages. The results used in a valuation would give the *relative* accuracy of the three methods. I have worked out the formula with the Z weights (H^M Table), and found that

Age at maturity = $57 \cdot 595 + \cdot 1200 \times$ (unexpired term)

The results could also be used as a rough check on the average ages at valuations, and there certainly seems a possibility of doing something towards making a simple "model office" for endowment assurances with the help of the method we have been using.

* The method used by me was approximate and can probably be improved; the result is merely given as an indication of a possible line for research.

CHAPTER VIII

THEORETICAL DISTRIBUTIONS.
SPURIOUS CORRELATION

1. In the previous chapter we saw that it was natural to want a scale for measuring correlation and we showed that if we simplify the full table by fitting a straight line* to the statistics, then its slope might be taken as a measure of correlation. But, though this seems reasonable on the evidence, it is not conclusive; it might be better to use some function of the slope of the line rather than the slope itself, or, we might find from experience that a straight line was not the best thing to use in our simplification. We have, therefore, to see if these doubts can be removed, and a way to do this is to consider correlation from a theoretical standpoint by building up tables in which we can estimate the amount of correlation from general considerations.

2. Various correlation tables can be devised, but we may begin by taking a case where ten coins are tossed and eight of them are left on the table, the other two being re-tossed. Then we have a pair of tossings in which eight coins out of ten are common to each member of the pair. We repeat the experiment a number of times and produce a correlation table in which, as eight out of ten coins are fixed, we may expect the correlation to be measured by ·8 or at any rate by a function of 8. Similarly, if we leave 5 coins the coefficient should be ·5 and if we leave 2 coins it should be ·2. The tables worked out theoretically would be as shown on pp. 165–7. These tables are symmetrical; the two standard deviations are the same and

---

* We reach two straight lines for each correlation table; one corresponding to the means of the columns and the other corresponding to the means of the rows.

the means (see last column and bottom row) run in a straight line and the slope of the line, judged by the tangent of the angle it makes with the horizontal, is ·8 or ·5 or ·2.

3. We may indicate how the tables were formed by taking the one that gives the numbers when eight coins are left. Consider those cases in which there were ten "heads" at the first throw. Then all the eight coins left must be "heads". The other two on being re-tossed will be in the following proportions:

2 Heads ...	...	...	1 case
1 Head and 1 Tail		...	2 cases
2 Tails ...	...	...	1 case

Thus, with the eight heads left, we conclude that for four cases producing 10 heads at the first throw, one will produce 10 heads at the second throw, two will produce 9 heads, and one will produce 8 heads. Now consider the next case where there are nine "heads" and one "tail" at the first throw. Then we can leave either eight heads or seven heads and one tail; the number of ways in which we can do this is $_9C_8$ and $_9C_7$, that is 9 and 36 or, as we are only concerned with proportions, as 1 : 4. The two re-tossed coins will be thrown in the proportion of $1HH : 2HT : 1TT$; and we can then produce the second column. The reader will appreciate that the totals of the columns will be a multiple of the terms of the binomial $(\frac{1}{2} + \frac{1}{2})^{10}$.

The coefficients worked out by the methods of Chapter VII give the values ·8, ·5 and ·2. For example $S(xy)$ for the ·8 table will be found to be 8192 and $\sigma_1 = \sigma_2 = \sqrt{2·5}$. Therefore

$$r = \frac{S(xy)}{N\sigma_1\sigma_2} = \frac{8192}{4096 \times 2·5} = ·8$$

4. Let us see if we can use these tables to help us to decide whether ·8 or a function of ·8 should be used as the measure or coefficient of correlation. An easy experiment is to add the ·8 table and the ·2 table together after increasing the former so that the two tables represent the distribution of the same total number of cases. The result of such a process is that the means

(157)

of the rows (or columns) will be half-way between those shown in the two tables from which the composite table is formed. These means are identical with those of the ·5 table, although the distribution of the cases is different. This is evidence that we can assume that ·8 or ·5 or ·2 is a proper coefficient of correlation and that we need not speculate with functions of these figures. If we generalise our result we may say that we want to find a function of r such that

$$f(r_1) + f(r_2) = f\left(\frac{r_1 + r_2}{2}\right)$$

and this is satisfied by writing $f(r) = r$.

5. It will, however, be noticed that we have chosen a particular case where the distribution is based on a symmetrical binomial and it does not follow that other cases will be so easy to interpret. We can, however, form similar tables with dice where we regard two of the six faces as "head" and four other faces as "tail". We then get distributions of the form $(\frac{1}{3} + \frac{2}{3})^{10}$, the means of the columns (or rows) are in a straight line and we reach the means of the ·5 table by adding together equally large tables giving correlation of ·8 and ·2. The $r = ·8$ table is given on p. 168. Admittedly we have even now only dealt with tables of double entry corresponding to frequency distributions like the binomial series and we cannot expect all the distributions that occur in practice to be so simple. We must not assume that in every case the means will follow a straight line nor are we entitled to say that the slope of the straight line will give a correct measure of correlation if the distribution diverges considerably from those discussed, but the large majority of correlation tables conform approximately to the type we have indicated.

6. The reader will have noticed that in the work we have just been doing we have dealt with a series of points analogous to the binomial series and not with a surface analogous to a frequency-curve. The normal curve with which we dealt in a previous chapter is, in certain conditions, the limit of the binomial series and the frequency surface* corresponding with the normal curve is

$$Z = \frac{N}{2\pi\sigma_1\sigma_2\sqrt{(1-r^2)}} e^{-\frac{1}{2}\left\{\frac{x^2}{\sigma_1^2(1-r^2)} - \frac{2xyr}{\sigma_1\sigma_2(1-r^2)} + \frac{y^2}{\sigma_2^2(1-r^2)}\right\}}$$

Now, if $r = 0$ this expression reduces to the product of two normal curves and if we find $\int_{-\infty}^{\infty}\int_{-\infty}^{\infty} Zxy\,dx\,dy$, we reach $Nr\sigma_1\sigma_2$ or $r = \dfrac{(xy)\ \text{moment}}{N\sigma_1\sigma_2}$ as we have already supposed in Chapter VII.

7. The normal surface has some properties to which special attention may be called. If we examine the distribution of an array of type t, we see that it is

$$Z = Z_0 e^{-\{g_1 x^2 - 2htx + g_2 t^2\}}$$

where g_1, h and g_2 are written for the longer expressions in σ_1, σ_2 and r.

Making the index a perfect square, we have

$$Z = Z_0 e^{-g_1\left\{x - \frac{ht}{g_1}\right\}^2} e^{-g_2 t^2 + \frac{h^2 t^2}{g_1}}$$

$$= Z_0' e^{-g_1\left\{x - \frac{ht}{g_1}\right\}^2}$$

which is a normal distribution having the same standard deviation as that of the whole surface, but its mean differs from that of the whole surface by ht/g_1. It follows that:

(1) the deviation of the mean of the array is directly proportional to the type; or, in other words, the means of arrays increase or decrease in arithmetical progression and so lie on a straight line,

(2) the standard deviations of all parallel arrays are equal and independent of their types.

So far as the former of these conclusions is concerned, we have the same property as that found in our coin-tossing tables and assumed in the previous chapter. The other property is not found with our coin-tossing tables. It must not be con-

* See Appendix III.

cluded from this that the normal surface has so small a scope as to be of little practical use; it has, probably, a far larger scope than the analogous normal curve has in frequency-curve work. It may help the reader to visualise the surface if he bears in mind the following points:

(1) vertical sections cut parallel to the axis of x or the axis of y are normal curves,

(2) the contour lines are ellipses and if these ellipses are projected on to the plane $Z = 0$ they are concentric, similar and similarly situated.

The appearance would be that of an isolated hill standing on a wide plain. This plain rises very slowly as we approach the hill, then the hillside becomes gradually steeper until, as we near the top, it becomes less steep and the top is nearly, but not quite, flat. The hill is narrowest when seen from the north-west or south-east and widest when seen from the north-east or south-west.

8. We may now discuss a danger against which we must be on guard in statistical work on correlation. The danger is that correlation may be revealed when it is absent, or exaggerated when present, in consequence of the arrangement of the statistical material. We will consider two causes of the introduction of this "spurious correlation". The first may be taken from our coin-tossing tables. We saw that by adding together the ·8 and ·2 tables in equal proportions we reached a table which gave a correlation of ·5. But let us see what would happen if we added together two tables where $r = $ ·5 but shifted the mean of one of them. This might happen in practical work if two persons, recording similar objects, measured correctly except that one always overstated his results by a constant figure. The results are then amalgamated and the table formed might then be similar to the table on p. 169. The coefficient of correlation is worked out and found to be ·78.

9. We may now consider how we might detect the cases in which this sort of thing happens. The means of the various rows run in a particular way; they begin and end as the $r = \cdot 5$ tables but, when the amalgamation comes in, the run of the line is such as to join the two end pieces together with a curved line. Again, the totals do not form a binomial or any single frequency-curve. In the particular case these two points would be sufficient warning, but in practice it is hard to apply them because the ends of an experience, being based on relatively small numbers, obscure the real shape of the regression lines and the curve formed by the totals. There may also be many observers instead of only two, and these observers might turn the end pieces into curved lines and give a regression line like a flattened S. The real remedy in such cases is to see that the various experiences grouped together are alike as regards both their means and their distributions and to use amalgamated figures only when the amalgamation is justified.

10. Another way in which a spurious correlation may be introduced arises through the use of indices. As an example we may refer to endowment assurances by limited payments on the books of a company doing a large quantity of such business and consider the term of the original assurance (t_1), the number of premiums to be paid in future (t_2), and the number of years for which the policy has been in force (t_3). If we formed the ratios t_2/t_1 and t_3/t_1, and worked out the coefficients of correlation, we should not obtain a measure of the correlation between number of premiums payable in future and the number of years in force because the result of using fractions with the same denominator in each would be to exaggerate correlation —that is, to introduce spurious correlation.

The general propositions of spurious correlation, of which the result just mentioned is a particular case, are as follows:

I. *To find the mean of an index in terms of the means, standard deviations and coefficient of correlation of the two absolute measurements.*

Let x_1, x_2, x_3, x_4 be the absolute sizes of any four correlated subjects; m_1, m_2, m_3, m_4 their mean values; σ_1, σ_2, σ_3, σ_4 their standard deviations; $r_{12}, r_{23}, r_{34}, r_{41}, r_{24}, r_{13}$ the six coefficients of correlation; $\epsilon_1, \epsilon_2, \epsilon_3, \epsilon_4$ the deviations of the four subjects from their means, i.e. $x_1 = m_1 + \epsilon_1$, etc.; i_{13} the mean value of the index x_1/x_3, and i_{24} the mean value of x_2/x_4; Σ_1 and Σ_2 the standard deviations of the indices x_1/x_3 and x_2/x_4 respectively, and N the total number of groups.

We shall suppose the ratios of the deviations from the mean values of the organs are so small that their cubes may be neglected. Then

$$i_{13} = \frac{1}{N} S\left(\frac{x_1}{x_3}\right) = \frac{1}{N} \cdot \frac{m_1}{m_3} ; S\left\{\left(1 + \frac{\epsilon_1}{m_1}\right)\left(1 + \frac{\epsilon_3}{m_3}\right)^{-1}\right\}$$

$$= \frac{1}{N} \cdot \frac{m_1}{m_3}\left\{N + \frac{S(\epsilon_1)}{m_1} - \frac{S(\epsilon_3)}{m_3} - \frac{S(\epsilon_1\epsilon_3)}{m_1 m_3} + \frac{S(\epsilon_3)^2}{m_3^2}\right\}$$

But $S(\epsilon_1) = S(\epsilon_3) = 0$ and $S(\epsilon_1\epsilon_3) = N\sigma_1\sigma_3 r_{13}$ and $S(\epsilon_3)^2 = N\sigma_3^2$

$$\therefore \qquad i_{13} = \frac{m_1}{m_3}\left(1 + \frac{\sigma_3^2}{m_3^2} - \frac{\sigma_1}{m_1}\cdot\frac{\sigma_3}{m_3}\cdot r_{13}\right)$$

and

$$i_{24} = \frac{m_2}{m_4}\left(1 + \frac{\sigma_4^2}{m_4^2} - \frac{\sigma_2}{m_2}\cdot\frac{\sigma_4}{m_4}\cdot r_{24}\right)$$

II. *To find the standard deviation of an index in terms of the standard deviations and coefficient of correlation of the two absolute measurements.*

$$N \times \Sigma_{13}^2 = S\left(\frac{x_1}{x_3} - i_{13}\right)^2$$

$$= \frac{m_1^2}{m_3^2} S\left\{\left(1 + \frac{\epsilon_1}{m_1}\right)\left(1 + \frac{\epsilon_3}{m_3}\right)^{-1} - \left(1 + \frac{\sigma_3^2}{m_3^2} - \frac{\sigma_1}{m_1}\cdot\frac{\sigma_3}{m_3}\cdot r_{13}\right)\right\}^2$$

$$= i_{13}^2 S\left\{\frac{\epsilon_1}{m_1} - \frac{\epsilon_3}{m_3} + \text{square terms}\right\}^2$$

$$= i_{13}^2\left(N\frac{\sigma_1^2}{m_1^2} + N\frac{\sigma_3^2}{m_3^2} - 2N\frac{\sigma_1}{m_1}\cdot\frac{\sigma_3}{m_3}\cdot r_{13}\right)$$

or $\qquad \Sigma_{13} = i_{13}\sqrt{\left\{\frac{\sigma_1^2}{m_1^2} + \frac{\sigma_3^2}{m_3^2} - 2\frac{\sigma_1}{m_1}\cdot\frac{\sigma_3}{m_3}\cdot r_{13}\right\}}$

III. *To find the coefficient of correlation of two indices in terms of the coefficients of correlation of four absolute measurements and their standard deviations.*

Let x_1/x_3 and x_2/x_4 be the two indices.

Then, if ρ be the coefficient of correlation of the two indices,

$$N\rho\Sigma_{13}\Sigma_{24} = S\left(\frac{x_1}{x_3} - i_{13}\right)\left(\frac{x_2}{x_4} - i_{24}\right)$$

$$= \frac{m_1 m_2}{m_3 m_4} S\left(1 + \frac{\epsilon_1}{m_1} - \frac{\epsilon_3}{m_3} - \frac{\epsilon_1\epsilon_3}{m_1 m_3} + \frac{\epsilon_3^2}{m_3^2} - 1 - \frac{\sigma_3^2}{m_3^2} + \frac{\sigma_1}{m_1}\cdot\frac{\sigma_3}{m_3}\cdot r_{13}\right)$$

$$\times \left(1 + \frac{\epsilon_2}{m_2} - \frac{\epsilon_4}{m_4} - \frac{\epsilon_2\epsilon_4}{m_2 m_4} + \frac{\epsilon_4^2}{m_4^2} - 1 - \frac{\sigma_4^2}{m_4^2} + \frac{\sigma_2}{m_2}\cdot\frac{\sigma_4}{m_4}\cdot r_{24}\right)$$

$$= i_{13}i_{24} S\left(\frac{\epsilon_1}{m_1} - \frac{\epsilon_3}{m_3}\right)\left(\frac{\epsilon_2}{m_2} - \frac{\epsilon_4}{m_4}\right)$$

as we neglect the terms of cubic order. Therefore

$$\rho\Sigma_{13}\Sigma_{24} = i_{13}i_{24}\left(\frac{\sigma_1}{m_1}\cdot\frac{\sigma_2}{m_2}r_{12} - \frac{\sigma_1}{m_1}\cdot\frac{\sigma_4}{m_4}r_{14} - \frac{\sigma_2}{m_2}\cdot\frac{\sigma_3}{m_3}r_{23} + \frac{\sigma_3}{m_3}\cdot\frac{\sigma_4}{m_4}r_{34}\right)$$

Hence

$$\rho = \frac{\dfrac{\sigma_1}{m_1}\cdot\dfrac{\sigma_2}{m_2}r_{12} - \dfrac{\sigma_1}{m_1}\cdot\dfrac{\sigma_4}{m_4}r_{14} - \dfrac{\sigma_2}{m_2}\cdot\dfrac{\sigma_3}{m_3}r_{23} + \dfrac{\sigma_3}{m_3}\cdot\dfrac{\sigma_4}{m_4}r_{34}}{\sqrt{\left(\dfrac{\sigma_1^2}{m_1^2} + \dfrac{\sigma_3^2}{m_3^2} - 2\dfrac{\sigma_1}{m_1}\cdot\dfrac{\sigma_3}{m_3}\cdot r_{13}\right)}\sqrt{\left(\dfrac{\sigma_2^2}{m_2^2} + \dfrac{\sigma_4^2}{m_4^2} - 2\dfrac{\sigma_2}{m_2}\cdot\dfrac{\sigma_4}{m_4}\cdot r_{24}\right)}}$$

Proposition I shows that the mean of an index is not the ratio of the means of the corresponding absolute measurements, and Proposition III shows that the ρ will vanish when the four subjects forming the indices are quite uncorrelated, while, if two, say, the third and fourth, are identical, so that $r_{34} = 1$ and $\sigma_3/m_3 = \sigma_4/m_4$, we have

$$\rho = \frac{\dfrac{\sigma_1}{m_1}\cdot\dfrac{\sigma_2}{m_2}r_{12} - \dfrac{\sigma_1}{m_1}\cdot\dfrac{\sigma_3}{m_3}r_{13} - \dfrac{\sigma_2}{m_2}\cdot\dfrac{\sigma_3}{m_3}r_{23} + \dfrac{\sigma_3^2}{m_3^2}}{\sqrt{\left(\dfrac{\sigma_1^2}{m_1^2} + \dfrac{\sigma_3^2}{m_3^2} - 2\dfrac{\sigma_1}{m_1}\cdot\dfrac{\sigma_3}{m_3}\cdot r_{13}\right)}\sqrt{\left(\dfrac{\sigma_2^2}{m_2^2} + \dfrac{\sigma_3^2}{m_3^2} - 2\dfrac{\sigma_2}{m_2}\cdot\dfrac{\sigma_3}{m_3}\cdot r_{23}\right)}}$$

This would become applicable in the case of endowment assurances by limited payments to which we referred.

11-2

An interesting special case arises when the subjects x_1, x_2, x_3 are not correlated and x_1/x_3 and x_2/x_3 are formed, then

$$\rho = \frac{\dfrac{\sigma_3^2}{m_3^2}}{\sqrt{\left(\dfrac{\sigma_1^2}{m_1^2} + \dfrac{\sigma_3^2}{m_3^2}\right)} \sqrt{\left(\dfrac{\sigma_2^2}{m_2^2} + \dfrac{\sigma_3^2}{m_3^2}\right)}}$$

11. The practical lessons about spurious correlation to be learnt from the foregoing are (1) to deal with homogeneous data and not to be too certain about the value of a coefficient in the case of amalgamated experiences until you are sure that those experiences are homogeneous; (2) to avoid making, or be careful in interpreting, correlation tables where the functions correlated are expressed as indices in which the denominators are identical or may themselves be correlated.

We may add that spurious correlation may arise when the correlated pairs relate to successive years, and so are not taken at random as regards time. If, however, the correlation between the two nth differences becomes equal to the correlation between the two $(n + 1)$th differences, we reach the correlation independent of time, provided the dependence of each variable on time takes the form $a + bt + bt^2 + \ldots$.

Coin-tossings with ten coins in pairs. Eight coins common to each member of pair

No. of heads in second tossing	No. of heads in first tossing											Total	Mean of row
	0	1	2	3	4	5	6	7	8	9	10		
0	1	2	1	…	…	…	…	…	…	…	…	4	1·0
1	2	12	18	8	…	…	…	…	…	…	…	40	1·8
2	1	18	61	72	28	…	…	…	…	…	…	180	2·6
3	…	8	72	176	168	56	…	…	…	…	…	480	3·4
4	…	…	28	168	322	252	70	…	…	…	…	840	4·2
5	…	…	…	56	252	392	252	56	…	…	…	1,008	5·0
6	…	…	…	…	70	252	322	168	28	…	…	840	5·8
7	…	…	…	…	…	56	168	176	72	8	…	480	6·6
8	…	…	…	…	…	…	28	72	61	18	1	180	7·4
9	…	…	…	…	…	…	…	8	18	12	2	40	8·2
10	…	…	…	…	…	…	…	…	1	2	1	4	9·0
Total	4	40	180	480	840	1,008	840	480	180	40	4	4,096	
Mean of column	1·0	1·8	2·6	3·4	4·2	5·0	5·8	6·6	7·4	8·2	9·0		

Five coins common to each member of pair

No. of heads in second tossing	No. of heads in first tossing											Total	Mean of row
	0	1	2	3	4	5	6	7	8	9	10		
0	1	5	10	10	5	1	:	:	:	:	:	32	2·5
1	5	30	75	100	75	30	5	:	:	:	:	320	3·0
2	10	75	235	400	400	235	75	10	:	:	:	1,440	3·5
3	10	100	400	860	1,100	860	400	100	10	:	:	3,840	4·0
4	5	75	400	1,100	1,780	1,780	1,100	400	75	5	:	6,720	4·5
5	1	30	235	860	1,780	2,252	1,780	860	235	30	1	8,064	5·0
6	:	5	75	400	1,100	1,780	1,780	1,100	400	75	5	6,720	5·5
7	:	:	10	100	400	860	1,100	860	400	100	10	3,840	6·0
8	:	:	:	10	75	235	400	400	235	75	10	1,440	6·5
9	:	:	:	:	5	30	75	100	75	30	5	320	7·0
10	:	:	:	:	:	1	5	10	10	5	1	32	7·5
Total	32	320	1,440	3,840	6,720	8,064	6,720	3,840	1,440	320	32	32,768	
Mean of column	2·5	3·0	3·5	4·0	4·5	5·0	5·5	6·0	6·5	7·0	7·5		

Two coins common to each member of pair

No. of heads in second tossing	No. of heads in first tossing											Total	Mean of row
	0	1	2	3	4	5	6	7	8	9	10		
0	1	8	28	56	70	56	28	8	1	...	...	256	4·0
1	8	66	240	504	672	588	336	120	24	2	...	2,560	4·2
2	28	240	913	2,024	2,884	2,744	1,750	728	184	24	1	11,520	4·4
3	56	504	2,024	4,768	7,280	7,504	5,264	2,464	728	120	8	30,720	4·6
4	70	672	2,884	7,280	11,956	13,328	10,192	5,264	1,750	336	28	53,760	4·8
5	56	588	2,744	7,504	13,328	16,072	13,328	7,504	2,744	588	56	64,512	5·0
6	28	336	1,750	5,264	10,192	13,328	11,956	7,280	2,884	672	70	53,760	5·2
7	8	120	728	2,464	5,264	7,504	7,280	4,768	2,024	504	56	30,720	5·4
8	1	24	184	728	1,750	2,744	2,884	2,024	913	240	28	11,520	5·6
9	...	2	24	120	336	588	672	504	240	66	8	2,560	5·8
10	...	...	1	8	28	56	70	56	28	8	1	256	6·0
Total	256	2,560	11,520	30,720	53,760	64,512	53,760	30,720	11,520	2,560	256	262,144	
Mean of column }	4·0	4·2	4·4	4·6	4·8	5·0	5·2	5·4	5·6	5·8	6·0		

Throwing ten dice in pairs. Eight dice common to each member of pair. Success is a "five" or "six"

No. of successes in second throw	No. of successes in first throw											Total	Mean of row
	0	1	2	3	4	5	6	7	8	9	10		
0	4,096	4,096	1,024	...	...	...	...	...	...	...	...	9,216	·6̇
1	4,096	20,480	17,408	4,096	...	...	...	...	...	...	...	46,080	1·4̇6̇
2	1,024	17,408	45,312	32,768	7,168	...	...	...	...	...	...	103,680	2·2̇6̇
3	...	4,096	32,768	58,368	35,840	7,168	...	...	...	...	...	138,240	3·0̇6̇
4	...	...	7,168	35,840	48,384	25,088	4,480	...	...	...	...	120,960	3·86
5	...	...	...	7,168	25,088	26,880	11,648	1,792	...	...	...	72,576	4·6
6	...	...	...	...	4,480	11,648	10,080	3,584	448	...	...	30,240	5·4̇6̇
7	...	...	...	...	...	1,792	3,584	2,496	704	64	...	8,640	6·2̇6̇
8	...	...	...	...	...	...	448	704	384	80	4	1,620	7·0̇6̇
9	...	...	...	...	...	...	...	64	80	32	4	180	7·8̇6̇
10	...	...	...	...	...	...	...	...	4	4	1	9	8·6
Total	9,216	46,080	103,680	138,240	120,960	72,576	30,240	8,640	1,620	180	9	531,441	
Mean of column	·6̇	1·4̇6̇	2·2̇6̇	3·0̇6̇	3·86	4·6	5·4̇6̇	6·2̇6̇	7·0̇6̇	7·8̇6̇	8·6		

Table formed by adding together two of the coin-tossing tables, one exactly like that on p. 166, and the other with the headings changed from 0–10 to 3–13

No. of successes in second record	No. of successes in first record														Total	Mean of row
	0	1	2	3	4	5	6	7	8	9	10	11	12	13		
0	1	5	10	10	5	1	...	...	...	...	...	...	...	...	32	2·5
1	5	30	75	100	75	30	5	...	...	...	...	...	...	...	320	3·0
2	10	75	235	400	400	235	75	10	...	...	...	...	...	...	1,440	3·5
3	10	100	400	861	1,105	870	410	105	11	...	...	...	...	...	3,872	4·012
4	5	75	400	1,105	1,810	1,855	1,200	475	105	10	...	...	...	...	7,040	4·568
5	1	30	235	870	1,855	2,487	2,180	1,260	470	105	11	...	...	...	9,504	5·226
6	...	5	75	410	1,200	2,180	2,640	2,200	1,260	475	105	10	...	...	10,560	6·045
7	...	...	10	105	475	1,260	2,200	2,640	2,180	1,200	410	75	5	...	10,560	6·955
8	...	...	...	11	105	470	1,260	2,180	2,487	1,855	870	235	30	1	9,504	7·774
9	...	...	...	...	10	105	475	1,200	1,855	1,810	1,105	400	75	5	7,040	8·432
10	...	...	...	...	...	11	105	410	870	1,105	861	400	100	10	3,872	8·988
11	...	...	...	...	...	...	10	75	235	400	400	235	75	10	1,440	9·5
12	...	...	...	...	...	...	...	5	30	75	100	75	30	5	320	10·0
13	...	...	...	...	...	...	...	...	1	5	10	10	5	1	32	10·5
Total	32	320	1,440	3,872	7,040	9,504	10,560	10,560	9,504	7,040	3,872	1,440	320	32	65,536	
Mean of columns	2·5	3·0	3·5	4·012	4·568	5·226	6·045	6·955	7·774	8·432	8·988	9·5	10·0	10·5		

CORRELATION OF CHARACTERS NOT QUANTITATIVELY MEASURABLE

1. Before the theory in this section is discussed we will give a table showing the class of problem with which it deals, drawn from vaccination statistics and relating to the Sheffield smallpox outbreak of 1887–8:*

Degree of effective vaccination	Strength to resist Smallpox when incurred			
	Cicatrix	Recoveries	Deaths	Total
Present ...		3,951	200	4,151
Absent ...		278	274	552
Total ...		4,229	474	4,703

The characters with which we are concerned are "Strength to resist smallpox when incurred" and "Degree of effective vaccination", and the statistics cannot be arranged in a more detailed manner. The characters cannot be measured quantitatively; but as the absence of such measurement does not mean that there is no correlation, we must see how the coefficient can be obtained in such a case.

2. Let us consider this problem in the first place by seeing if we can write down a few cases in which we can assign a value to the coefficient of correlation from general considerations. If we toss a coin, it must come down "head" or "tail"; if we form pairs as in Chapter VIII, by pairing consecutive tossings, there will be no correlation and in a table such as that of the previous

---

* *Biometrika*, I, 375 et seq. This paper, by W. R. Macdonell, and a supplementary one deal with the subject in a way that shows clearly the strength of the evidence on the side of vaccination. The question of class is investigated, a practical point frequently neglected.

paragraph there would be an equal number in each division. But if we made a pair by leaving the coin on the table, and counting it a second time, we should have absolute correlation and we should have in our table an equal number in the top left-hand and bottom right-hand divisions, the other two divisions being blank. If we amalgamate these two tables, assuming that the total of each is 4, we reach the table shown below, having a coefficient of correlation of ·5.

Second tossing	First tossing		Total
	Head	Tail	
Head	3	1	4
Tail	1	3	4
Total	4	4	8

In these simple cases, where the four divisions represent the frequencies at four separate points, the correct value of r is given by the expression

$$\frac{ad - bc}{\sqrt{\{(a+b)(c+d)(a+c)(b+d)\}}}$$

where a, b, c and d have the meanings indicated in the scheme of §5.

3. It does not, however, follow that this expression is one which may be used in all circumstances as, though there are only four divisions, the things measured may imply a continuous scale of measurement even though we cannot or do not express it in detailed fashion. Thus, in our vaccination statistics, the degree of successful vaccination may vary between a vaccination in infancy, for a person aged 40 at the time of the epidemic, and a series of vaccinations, the last of which has been recently performed. Again, the power of recovery when attacked may also be deemed to lie on a longer scale than that implied by the two divisions "recoveries" and "deaths". To take another example we could, if we were studying eye-colour in parent and offspring, make a scale of colour from black down to pale blue (or to absence of pigment in albinotic cases), but the statistics might be available merely in the form "brown" and "not brown". In other words the statistics in a four-fold correlation table may relate to a continuous frequency distribution like the table on p. 142, but owing to the way the facts had to be stated or collected there are only four divisions for the whole of the material. Our first problem is to see whether the simple formula at the end of §2 will give a satisfactory answer in these circumstances and if it fails what alternative may be adopted.

4. In Chapter VIII we gave tables based on coin-tossing and we might group the material of one of these tables into four divisions and see what answer the formula in question gives. If we take the table having a correlation of ·5 and cut it between the 5 "heads" and 6 "heads", we reach the following:

Number of heads in second tossing	NUMBER OF HEADS IN FIRST TOSSING		Total
	0–5	6–10	
0–5	15,330	5,086	20,416
6–10	5,086	7,266	12,352
Total	20,416	12,352	32,768

The formula gives

$$\frac{15330 \times 7266 - 5086 \times 5086}{20416 \times 12352} = ·34$$

which is far removed from the true value of ·5.

5. It is clear from this evidence that we must look for another solution and having seen in the previous chapter that we could express a frequency surface as

$$Z = \frac{N}{2\pi \sqrt{(1-r^2)}\,\sigma_1\sigma_2} e^{-\frac{1}{2}\frac{1}{1-r^2}\left(\frac{x^2}{\sigma_1^2}+\frac{y^2}{\sigma_2^2}-\frac{2rxy}{\sigma_1\sigma_2}\right)}$$

we may now consider what conclusions we may draw if we divide this surface into four parts by two planes at right angles to the axes of x and y at distances h' and k' from the origin, as suggested by the figures on p. 173.

$$y$$

$$a \qquad\qquad b$$

$$k'$$

$$-x \qquad\qquad \ggg h' \longrightarrow \qquad\qquad x$$

$$c \qquad\qquad d$$

$$-y$$

Table of Frequencies

a	b	$a+b$
c	d	$c+d$
$a+c$	$b+d$	N

Then

$$d = \frac{N}{2\pi\sqrt{(1-r^2)}\,\sigma_1\sigma_2}\int_{h'}^{\infty}\int_{k'}^{\infty} e^{-\frac{1}{2}\frac{1}{1-r^2}\left(\frac{x^2}{\sigma_1^2}+\frac{y^2}{\sigma_2^2}-\frac{2rxy}{\sigma_1\sigma_2}\right)}\,dx\,dy$$

$$= \frac{N}{2\pi\sqrt{(1-r^2)}}\int_{h}^{\infty}\int_{k}^{\infty} e^{-\frac{1}{2}\frac{1}{1-r^2}(x^2+y^2-2rxy)}\,dx\,dy$$

by substituting x^2 for $\dfrac{x^2}{\sigma_1^2}$ and y^2 for $\dfrac{y^2}{\sigma_2^2}$

and writing $\qquad h = \dfrac{h'}{\sigma_1}$ and $k = \dfrac{k'}{\sigma_2}$

Further $\qquad b+d = \dfrac{N}{\sqrt{(2\pi)}\,\sigma_1}\displaystyle\int_{h'}^{\infty} e^{-\frac{1}{2}\frac{x^2}{\sigma_1^2}}\,dx$

$$= \frac{N}{\sqrt{(2\pi)}}\int_{h}^{\infty} e^{-\frac{1}{2}x^2}\,dx$$

and $\qquad c+d = \dfrac{N}{\sqrt{(2\pi)}}\displaystyle\int_{k}^{\infty} e^{-\frac{1}{2}y^2}\,dy$

and, remembering that N the total frequency $= a+b+c+d$, we have

$$N - 2(b+d) = N - N\sqrt{\frac{2}{\pi}}\int_{h}^{\infty} e^{-\frac{1}{2}x^2}\,dx$$

$$\therefore \qquad \frac{(a+c)-(b+d)}{N} = \sqrt{\frac{2}{\pi}}\int_{0}^{h} e^{-\frac{1}{2}x^2}\,dx$$

and, similarly,

$$\frac{(a+b)-(c+d)}{N} = \sqrt{\frac{2}{\pi}}\int_{0}^{k} e^{-\frac{1}{2}y^2}\,dy$$

As a, b, c, and d are known, h and k can be found from Sheppard's Tables, and the problem becomes

"To find a value for r from the equation

$$\frac{N}{2\pi\sqrt{(1-r^2)}}\int_{h}^{\infty}\int_{k}^{\infty} e^{-\frac{1}{2}\frac{1}{1-r^2}(x^2+y^2-2rxy)}\,dx\,dy = d$$

where d, N, h, and k are known."

(174)

The solution (see Appendix IV) leads to the following equation:

$$\frac{ad-bc}{N^2HK} = r + \frac{r^2}{2}hk + \frac{r^3}{6}(h^2-1)(k^2-1) + \frac{r^4}{24}h(h^2-3)k(k^2-3)$$

$$+ \frac{r^5}{120}(h^4-6h^2+3)(k^4-6k^2+3)$$

$$+ \frac{r^6}{720}h(h^4-10h^2+15)k(k^4-10k^2+15)$$

$$+ \frac{r^7}{5040}(h^6-15h^4+45h^2-15)$$

$$\times (k^6-15k^4+45k^2-15) + \text{etc.}$$

where $\quad H = \dfrac{1}{\sqrt{(2\pi)}}e^{-\frac{1}{2}h^2}$ and $K = \dfrac{1}{\sqrt{(2\pi)}}e^{-\frac{1}{2}k^2}$

The numerical solution has to be obtained by approximating to the roots, and Newton's method* is convenient for the purpose.

6. The numerical work of our first example is as follows:

$$\sqrt{\frac{2}{\pi}}\int_0^h e^{-\frac{1}{2}x^2}dx = \frac{(a+c)-(b+d)}{N} = \frac{3755}{4703}$$

$$= \cdot 7984265$$

$$\therefore \qquad\qquad h = 1\cdot 27716$$

by interpolation in Sheppard's Tables (see *Tables for Statisticians*). In using these tables for this purpose, remember that the value ·7984265 corresponds to α in his notation, so

$$\tfrac{1}{2}(1 + \cdot 7984265) = \cdot 8992132$$

must be looked up inversely in his Table II. If his Table III be used, it must be entered with ·7984265.

---

* *Newton's method of approximating to the root of an equation.* Let $f(x) = 0$ be an equation from which the value of x is to be found and let b be a value near to x so that $x = b + h$ where h is small, then $f(x) = f(b+h) = f(b) + hf'(b) +$ terms involving higher powers of h by Taylor's Theorem, and since $f(x) = 0$, we have $h = -\dfrac{f(b)}{f'(b)}$ or $x = b - \dfrac{f(b)}{f'(b)}$. The chief objection to the method is that there may be more than one root near the value b, but this does not hold in the application to correlation. (Cf. Approximations to rate of interest from an annuity, Todhunter's *Interest and Annuities Certain*, p. 177, formula 2.)

Similarly

$$\sqrt{\frac{2}{\pi}} \int_0^k e^{-\frac{1}{2}v^2} dy = \cdot 7652561$$

$$\therefore \qquad k = 1 \cdot 18833$$

We next require $\dfrac{ad-bc}{N^2 H K}$; and we first get from Sheppard's Tables

$$H = \cdot 1764870 \qquad \therefore \quad \log H = \bar{1} \cdot 2467127$$

$$K = \cdot 1969111 \qquad \therefore \quad \log K = \bar{1} \cdot 2942702$$

Hence $\qquad\qquad \log \dfrac{ad-bc}{N^2 H K} = \cdot 1258266$

and $\qquad\qquad \dfrac{ad-bc}{N^2 H K} = 1 \cdot 336062$

Dr Macdonell gives 56 instead of 62 as the last two figures; the difference is probably due to interpolation.

Turning to the expression for r, we notice that hk is a product in the coefficients of r^2, r^4, r^6, etc., so it is well to work out its value and keep a note of it while the coefficients are being found. It is also advisable to begin the work by writing down the first six or seven powers of h and k.

Macdonell gives the following series:

$$\cdot 097083 r^7 + \cdot 008170 r^6 + \cdot 119614 r^5 + \cdot 137450 r^4$$
$$+ \cdot 043352 r^3 + \cdot 758844 r^2 + r = 1 \cdot 336056$$

In order to obtain r we must find a value near the true one as a first approximation.

Taking $\qquad\qquad \cdot 758844 r^2 + r - 1 \cdot 336056 = 0$

we have $\qquad r = \dfrac{-1 + \sqrt{\{1 + 4 \times 1 \cdot 336 \times \cdot 7588\}}}{1 \cdot 5177}$

$$= \cdot 8$$

Now, this value will be in excess of the truth because we have used only two terms of the series on the left-hand side of the equation for finding r, and we may take $\cdot77$ as a trial rate. Applying Newton's Rule, we have:

$$r = \cdot77 - \dfrac{\begin{array}{l} -1\cdot336056 + (\cdot77) + \cdot7588(\cdot77)^2 + \cdot0434(\cdot77)^3 \\ + \cdot1375(\cdot77)^4 + \cdot1196(\cdot77)^5 + \cdot0082(\cdot77)^6 + \cdot0971(\cdot77)^7 \end{array}}{\begin{array}{l} 1 + 2(\cdot77)(\cdot7588) + 3(\cdot77)^2(\cdot0434) + 4(\cdot77)^3(\cdot1375) \\ + 5(\cdot77)^4(\cdot1196) + 6(\cdot77)^5(\cdot0082) + 7(\cdot77)^6(\cdot0971) \end{array}}$$

$$= \cdot77 - \frac{\cdot0022}{2\cdot861}$$

$$= \cdot7692$$

In work such as this a table giving the first seven powers of the natural numbers is a help.

7. Tables of various functions required for the arithmetical work will be found in *Tables for Statisticians*. The term "tetrachoric functions"* is employed there. These tables are arranged so that we can use the equation

$$d/N = \tau_0\tau_0' + \tau_1\tau_1' r + \tau_2\tau_2' r^2, \text{ etc.}$$

where the values of τ are tabulated up to τ_{19}, and further values can be obtained by a difference formula given in the introduction to the tables.

The calculation of the coefficient has been set out above in detail, but with the help of Tables VIII and IX of *Tables for Statisticians*, Part II, much of the work can be avoided. All that has to be done, if these tables are available, is (1) to calculate h and k as shown in §6, (2) to calculate the ratio that the number in quadrant d bears to the total number of cases, i.e. d/N, and (3) to interpolate in the tables so as to obtain r.

8. We may now return to our coin-tossing, and we find that if we work out the coefficient of correlation for the table in §4

---

* The tetrachoric functions are closely allied to the Hermite polynomials and provide the fullest tables available. The sth tetrachoric function is (cp. p. 130)

$$\tau_s(h) = \frac{(-1)^{s-1}}{\sqrt{s!}} \frac{d^{s-1}}{dh^{s-1}} \left(\frac{1}{\sqrt{(2\pi)}} e^{-\frac{1}{2}h^2} \right)$$

and the $(s-1)$th Hermite polynomial is

$$H_{s-1}(h) = \frac{(-1)^{s-1}}{\frac{1}{\sqrt{(2\pi)}} e^{-\frac{1}{2}h^2}} \frac{d^{s-1}}{dh^{s-1}} \left(\frac{1}{\sqrt{(2\pi)}} e^{-\frac{1}{2}h^2} \right)$$

by the method just discussed, we reach a value of between ·51 and ·52. This is a good result, especially when we remember that the coin-tossing is not an absolutely continuous scale.

The broad conclusion that may be reached is that the assumptions lead to reasonable results in the kind of cases we have tested. The method does not work so well when the frequency surface is cut far from the mean and the numerical results in such cases should not be assumed to have minute accuracy.

9. We have assumed that the data available are only divided into four divisions and we shall postpone till later (Chapter XII) the discussion of correlation when the characteristics are not quantitatively measurable but are divided into several categories. We may, however, now deal with the case in which one variate is and the other is not quantitatively measurable as, for instance, in the table on p. 179 relating to the effect of enlarged glands on the weight of children (boys).* Though the statistics are divided into good and bad glands, the condition of glands is a continuous variate: some of the boys with bad glands were worse than others.

If the reader considers a volume of frequency built out of a complete table such as that for endowment assurances, or out of a correlation table giving relative ages of husbands and wives, he will see that he has a complete distribution. Now, if a volume of frequency be cut off from such a complete volume by a vertical plane at a given value of one variate, then the vertical through the centroid of this volume cuts the regression line. The vertical plane in the two-row table is at the division of the rows; in our example where the good glands end and the bad glands begin. If $\bar{p}$ and $\bar{q}$ be the co-ordinates of the point of section where the vertical through the centroid of the volume cuts the regression line, then we have, σ_1 and σ_2 being the standard deviations of the two variates and r the correlation,

$$\bar{p} = r\frac{\sigma_1}{\sigma_2}\bar{q} \qquad \text{or} \qquad r = \frac{\bar{p}}{\sigma_1}\bigg/\frac{\bar{q}}{\sigma_2}$$

* The method is given by K. Pearson in *Biometrika*, VII, 96 et seq., and the example is taken from that paper.

Weight	Boys with good glands	Boys with bad glands	Total
14	2	...	2
16	3	5	8
18	15	26	41
20	20	40	60
22	28	47	75
24	34	30	64
26	30	31	61
28	29	20	49
30	30	30	60
32	21	14	35
34	18	11	29
36	18	5	23
38	6	7	13
40	5	2	7
42	7	3	10
44	1	...	1
46	...	...	...
48	3	...	3
50	1	...	1
52	1	...	1
...	...	...	...
62	2	...	2
Total	274	271	545

Now $\bar{p}$ is the mean value of the quantitatively measurable variate for all the pairs with a certain one of the alternative variates, in our example, the mean weight of boys with bad glands and σ_1 is the standard deviation of all the boys. We cannot calculate $\bar{q}$ and σ_2 in a similar way, because they relate to glands of which no quantitative measure is available. If we assume the non-measurable variate (glands) to follow the normal probability distribution, the proportion of the non-measurable variate gives, with the help of tables of the probability integral, the ratio of y/σ_2 for the distance from the mean at which the division of this variate occurs, and then

$$\frac{\bar{q}}{\sigma_2} = \frac{N}{\sqrt{(2\pi)}\,\sigma_2} \int_y^\infty y e^{-\frac{1}{2}\frac{y^2}{\sigma_2^2}}\,dy \bigg/ \frac{N}{\sqrt{(2\pi)}\,\sigma_2} \int_y^\infty e^{-\frac{1}{2}\frac{y^2}{\sigma_2^2}}\,dy$$

$$= \frac{1}{\sqrt{(2\pi)}} e^{-\frac{1}{2}(y/\sigma_2)^2} \bigg/ \frac{1}{\sqrt{(2\pi)}} \int_{y/\sigma_2}^\infty e^{-\frac{1}{2}v^2}\,dy$$

The numerator is z and the denominator $\frac{1}{2}(1-\alpha)$ in the notation used in Sheppard's tables of the probability integral.

12-2

The working of the numerical example may help to make the method clearer; it is as follows:

The mean weight of all the boys is ... ... 27·7522

The standard deviation is ... ... 6·7502

The mean weight of boys with bad glands is 27·3737

$\frac{1}{2}(1-\alpha) = 271/545 = ·4972$

$\frac{1}{2}(1+\alpha) = ·5028$ and this value corresponds with

$z = ·3989$ in Sheppard's tables

The correlation of glands and weight is

$$\frac{(27·7522 - 27·3737)}{6·7502} \bigg/ \frac{·3989}{·4972}$$
$$= ·070$$

It may be remarked that the use of $\frac{1}{2}(1-\alpha)$ assumes that the column with the smaller total frequency will be taken; thus, in our example, there are fewer boys with bad glands than with good glands.

10. This example suggests a practical point, namely, that, before actually working out a coefficient of correlation, it is advisable to look at the statistics and form a preliminary idea of whether there is any correlation. In tables such as that on p. 142 there is no correlation if all the means of the rows are alike and all the means of the columns are alike. Similarly, in tables, such as the one on p. 170, if the entries within the table are proportional to the totals there is no correlation. In the example in § 9 above a comparison of the two inner columns with the total column shows that if there is any correlation it must be small because the distribution in the total would give a possible "graduation" of each of the inner columns.

CHAPTER X

STANDARD ERRORS

1. In statistical work we calculate a mean from a number of measurements, and we may be tempted to think that our work has definitely established the mean with which we are concerned. The arithmetical work may be correct in every detail and the measurements may have been made accurately, but the mean found from the statistics may differ from the true mean of the character measured because the things measured are limited in number—because, in other words, the sample we have taken does not exactly represent the unlimited population from which it is drawn. If we toss five coins and record the number of heads, we should obtain a table like the following where we give results of 140 repetitions of the experiment.*

Number of "heads" in trial	Number of trials in which the number of heads in previous column was recorded
5	4
4	24
3	49
2	40
1	20
0	3
	140

Now, treating this as a mere statistical problem, in which we do not know *a priori* anything about the true distribution, we may work out the mean number of "heads" as 2·6. This does not prove that this mean and no other can arise; a second

* Such an experiment may, alternatively, be regarded as drawing a random sample of 140 cases from an infinite population distributed as $(\frac{1}{2} + \frac{1}{2})^5$.

experiment might give a different result and we cannot, therefore, say from our calculation what the mean value really is. We can, however, approach the problem in another way; we can try to decide how deviations from a true mean are likely to be distributed and so form an opinion as to how a mean calculated from an experiment or series of experiments will differ from the truth. For practical purposes we might rest content if we could say that the true mean will not differ from the calculated mean by more than a small quantity (ϵ) once in a hundred trials. Before we go into the measures actually used, let us consider some of the points in a preliminary way.

All our statistical experience makes us feel sure that an experiment based on 1,000 cases must be more reliable than a similar experiment based on 50 cases, so we anticipate that ϵ, the small difference between the true and the calculated mean, will depend in some way on the number of cases. Again, a distribution that spreads widely gives the mean more opportunity to deviate than a distribution that is concentrated; so that we may also anticipate that ϵ will depend on the spread of the distribution, that is, on its standard deviation.

2. These remarks apply to all statistical measures. The measures are inexact and only approximate to the truth, but we can say that it is highly probable that they do not differ by more than a certain amount from the result which would be obtained if we could deal with an unlimited number of facts. In our discussion we have spoken of means, but every other measure is subject to the same general considerations and we must, therefore, consider what sort of value may be assigned to the small error ϵ for means, standard deviations, coefficients of correlation and other measures. We have anticipated that this error will depend on a standard deviation; and this sometimes leads to a little confusion because we must make up our minds as to the distribution to which the standard deviation refers. Let us imagine that we have worked out a coefficient of correlation for 100 pairs of, say, ages at marriage of husband and wife. Then we work out a second coefficient of correlation

for another hundred pairs and go on till we have a large number of these results. The coefficients will fall into a distribution like one of the frequency distributions we discussed in earlier chapters and it is the standard deviation of that distribution from its mean with which we are concerned. Similarly, we can repeat the coin-tossing experiment of § 1 over and over again and calculate the mean from each experiment. We may obtain any value for the mean between 5 and 0 heads; these extremes will only arise in the most unlikely case when every trial gave 5 heads or every trial gave no head. The most likely mean is 2·5 and if we repeat the experiment sufficiently we shall form an idea of the way the means are distributed: we shall reach a frequency distribution of means having its own mean, standard deviation, etc., and we must not confuse it with the frequency distributions such as that in the table in § 1 from which the means were calculated. It will help to avoid confusion of ideas if we speak of "standard error" when we are referring to the frequency distribution of a statistical measure (such as a mean, coefficient of correlation, etc.) instead of speaking of "standard deviation". The standard error is, then, the standard deviation of the frequency distribution of the particular measure we are examining.

3. With this introduction we may now consider the simplest of the bell-shaped frequency curves, namely, the normal curve of error, and see what conclusions we may draw if the distribution of a statistical measure takes that form. It has, in fact, been shown to be the form that the distribution of statistical measures tends to assume when the number of cases in the sample is large. Thus, even if the distribution in § 1 had been skew instead of symmetrical, the distribution of the means would have been more nearly of the form of the normal curve than the skew distributions from which they were obtained. By reference to the tables of this curve we see that the area corresponding to the standard deviation is about two-thirds of the whole area, while the area corresponding to twice the standard deviation is ·9545 of the whole area.

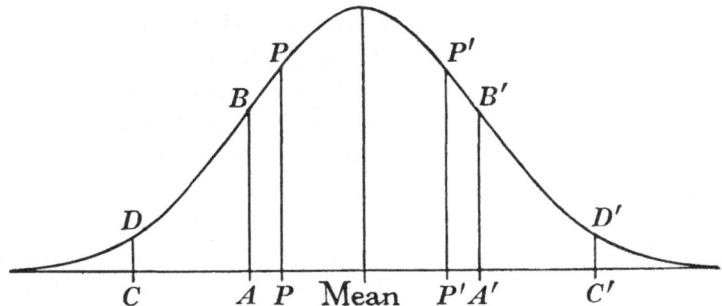

In other words, if the distribution takes this form we can say than an error of more than twice the standard error will occur 9 times in 200 trials and is, therefore, unlikely to have arisen in the particular case with which we are dealing. The diagram will help the reader to follow this argument. The area between AB which is at a distance equal to the standard deviation from the mean one side, and $A'B'$ at the same distance the other side, of the mean, is approximately two-thirds of the whole area. The lines CD and $C'D'$, which are twice as far from the mean, include nearly the whole curve; the pieces beyond those lines are tails which must be of relatively small dimensions.

4. It was formerly the custom to use another function known as the probable error, which is ·67449 times the standard error. The probable error gives that value of x (say p) which divides the part of the normal curve representing positive errors into two equal portions; it is therefore given by

$$\int_0^p \frac{1}{\sqrt{(2\pi)}}\, e^{-\frac{1}{2}x^2}\, dx = \cdot 25$$

where the whole area of the curve (positive *plus* negative deviations) is unity. In order to find p in terms of the standard deviation, we have, therefore, to obtain the value of x, corresponding to $\frac{1}{2}(1+\alpha) = \cdot 75$ in *Tables for Statisticians*, Part I, Table II or short table in Appendix IX, where α is

$$\int_{-x}^x \frac{1}{\sqrt{(2\pi)}}\, e^{-\frac{1}{2}x^2}\, dx$$

(184)

This can be done by interpolating inversely and p is thus found to be ·67449 approximately. The mean, or rather the vertical through the mean, divides the whole distribution into two equal parts: the probable error divides it into fourths and gives what Galton called the quartiles. The position is shown with the letters P and P' in the diagram; thrice the probable error includes about the same area as twice the standard error.

We may set down the following general rules:

(1) the true value and a calculated value of a mean or other characteristic are unlikely to differ by more than twice the standard error,

(2) if an experiment on any subject leads to a result which differs from that expected by more than twice the standard error we must suspect that we are not dealing with a random sample.

5. The problem before us is to consider how statistical measures calculated from limited data may vary about the expected values. Two methods of approach are available. We may, as indicated in §2, make a large number of experiments —or collect a large number of samples—and calculate the statistical measure in question for each of them. The procedure would generally be much too laborious, and we take, therefore, the second line of approach. Algebraic analysis based on the theory of probability enables us to determine the standard error that we should find in the limit if the sampling process were repeated indefinitely so that all possible samples were included in their expected proportions. We can often go further and determine the actual curve to which the frequency distribution of a particular statistical measure will tend as the number of samples is increased.

We may now take a simple illustration and find the standard error of the frequency, say n, with which an·event will happen in m independent trials where p is the probability of it happening and q of it failing. The probability of n being equal to m, $m-1$, ..., 2, 1, 0 is given by the terms of the binomial expansion. Taking moments about the point represented by p^m, the first moment is

$$mp^{m-1}q + m(m-1)p^{m-2}q^2 + \ldots + mq^m = mq(p+q)^{m-1} = mq.$$

The second moment about the same point is

$$mp^{m-1}q + 2m(m-1)p^{m-2}q^2 + \tfrac{3}{2}m(m-1)(m-2)p^{m-3}q^3$$
$$+ \ldots + m^2 q^m$$
$$= mp^{m-1}q + m(m-1)p^{m-2}q^2 + \frac{m(m-1)(m-2)}{2!}p^{m-3}q^3$$
$$+ \ldots + mq^m + m(m-1)p^{m-2}q^2 + m(m-1)(m-2)p^{m-3}q^3$$
$$+ \ldots + m(m-1)q^m$$
$$= mq + m(m-1)q^2$$

The second moment about the mean is, therefore,

$$mq + m(m-1)q^2 - m^2q^2 = mpq$$

∴ the standard error $= \sqrt{\mu_2} = \sqrt{(mpq)}$

That is to say, if we repeatedly make m independent trials the observed frequency of occurrence, n, will vary about the expected value mp with a standard error of $\sqrt{(mpq)}$.

6. We may now apply this result to a few examples:

(a) It has been remarked that the number of male children born is to the number of female children born as $1{,}050 : 1{,}000$; in other words, the probability of a child being male is $1{,}050/2{,}050$. If $51{,}350$ out of $100{,}000$ children proved to be males in a certain community, would it be safe to base on the statistics any theory connected with the variation from the usual probability? The expected result is $51{,}220$, and the standard error is

$$\sqrt{\left(100{,}000 \cdot \frac{1050}{2050} \cdot \frac{1000}{2050}\right)} = \pm\, 158 \cdot 07$$

The difference between the actual case and the expected result was 130, and as this is less than the standard error, no definite conclusion can be based on the divergence from the result.

(b) If the number of cases had been $10{,}000{,}000$, and the actual number $5{,}135{,}000$, then the standard error being

1,580·7 and the actual difference 13,000, it would have been sufficient evidence for the conclusion that the ratio 1,050 : 1,000 did not fit the particular case.

(c) If the probability of death within a year is ·007, the probable error in 200 cases is ·67449 $\sqrt{(200 \times \cdot 007 \times \cdot 993)}$ = 80, and it would, therefore, be possible to approximate to a loading for emergencies if 2·2 was taken instead of 1·4 as the number of deaths expected in a year out of 200 cases on risk for a year. The probable error would, I think, be preferable to the standard error for this purpose. That is, it would not be unreasonable to treat ·011 as the rate of mortality instead of ·007 in order to obtain some idea of an emergency loading for term assurances on the assumption that the number of cases is about 200 and the average age is such that ·007 might be taken as the probability of death in a year. It has also been assumed that it is correct to treat each class as if it were subject to its own rate of mortality and had to be treated independently of the rest of the business; that is, however, a debatable point.

(d) It will be noticed that if m remains constant, then $\sqrt{(mpq)}$ has its largest numerical value when $p = q = \frac{1}{2}$, which shows that an insurance office will generally find that if it has two classes of equal size, and one is subject to a higher rate of mortality than the other, the former will have the larger actual deviations from the expected number of claims, because the probability of dying in a year only reaches the value $\frac{1}{2}$ at the end of the mortality table.

7. We may now consider a frequency distribution divided into k groups such that the proportion of cases in the sth group is p_s and, clearly, $p_1 + p_2 + \ldots + p_s + \ldots + p_k = 1$. If we take a case at random from this distribution, the chance that it comes from the sth group is p_s and the chance that it comes from some other group is $q_s = 1 - p_s$. Let us suppose that m cases are taken at random and that n_s of them fall in the sth group. Then, though the expected value of n_s is mp_s, this frequency may assume values $m, m-1, \ldots, 2, 1, 0$ with probabilities given by

the terms of the binomial $(p_s + q_s)^m$ and the standard error of n_s will be

$$\sigma_{ns} = \sqrt{\{mp_s(1 - p_s)\}} \qquad \dots\dots(1).$$

If, in practice, we do not know the exact form of the frequency distribution from which the sample has been taken, we may approximate to the standard error by putting $p_s = n_s/m$, the observed proportion in the sth group of the sample. Hence, we have, approximately,

$$\sigma_{ns} = \sqrt{\{n_s(1 - n_s/m)\}} \qquad \dots\dots(2).$$

8. As the total of all the frequencies n_1, n_2, ... n_k is m, it follows that, if in a particular sample n_s is much greater than mp_s, the other frequencies must on the average be too small and this shows that the errors between the groups are correlated. The next point to be investigated is the amount of *the correlation between deviations in the frequencies of the sth and tth groups.*

The deviation, δn_s, of n_s from its expected value is $n_s - mp_s$.

As we are considering the relation between deviations in n_s and n_t we may conveniently class together all the remaining $k - 2$ frequency groups into a single remainder group, say, n_R. Then

$$n_s + n_t + n_R = m$$

$$p_s + p_t + p_R = 1$$

$$\delta n_s + \delta n_t + \delta n_R = 0$$

and $\qquad (\delta n_s + \delta n_t)^2 = (-\delta n_R)^2$

or $\qquad \delta n_s \delta n_t = \tfrac{1}{2}(\delta n_R^2 - \delta n_s^2 - \delta n_t^2)$

If we now imagine that a very large number, N, of random samples is taken and the expressions on both sides of the last equation are summed and divided by their number, N, then

$$\frac{1}{N} S(\delta n_s \delta n_t) = \frac{1}{2} \left\{ \frac{1}{N} S(\delta n_R^2) - \frac{1}{N} S(\delta n_s^2) - \frac{1}{N} S(\delta n_t^2) \right\}$$

The expressions on the right-hand side represent, in the limit, the squared standard errors of the group frequencies, given in equation (1) above. Hence in the limit

$$\frac{1}{N} S(\delta n_s \delta n_t) = \tfrac{1}{2}m\{p_R(1-p_R) - p_s(1-p_s) - p_t(1-p_t)\}$$
$$= \tfrac{1}{2}m\{(1-p_s-p_t)(p_s+p_t) - p_s(1-p_s) - p_t(1-p_t)\}$$
$$= -mp_s p_t \qquad\qquad\qquad ......(3).$$

But the correlation between n_s and n_t is

$$r_{n_s n_t} = \frac{\text{Limit of } \dfrac{1}{N} S(\delta n_s \delta n_t)}{\sigma_{n_s}\sigma_{n_t}}$$
$$= -\frac{p_s p_t}{\sqrt{\{p_s(1-p_s)\,p_t(1-p_t)\}}}$$
$$= -\sqrt{\left\{\frac{p_s p_t}{(1-p_s)(1-p_t)}\right\}} \qquad ......(4).$$

We may again approximate to this expression by substituting for p_s and p_t the proportionate frequencies, n_s/m and n_t/m, of the sample.

9. *To find the standard error of the mean of a sample of m observations.*

Let us again assume a frequency distribution divided into k groups where x_s is the value of the variable quantity x associated with the sth group. For the reasons already explained in the earlier Sections of this Chapter we must distinguish between (i) the mean of the population represented by the frequency distribution, namely

$$\bar{X} = \Sigma(p_s x_s)$$

where Σ indicates summation for all the k groups, and (ii) the mean calculated from a particular sample of m cases drawn at random from this population, namely

$$\bar{x} = \Sigma\left(\frac{n_s}{m}.x_s\right)$$

The standard error of $\bar{x}$, say $\sigma_{\bar{x}}$, will provide a measure of the extent to which the mean of the sample may differ from the mean of the population. The value of $\sigma_{\bar{x}}$ may be found by using the results (1) and (3) of the preceding sections.

Using a similar notation, we have

$$\delta\bar{x} = \bar{x} - \bar{X}$$

$$= \frac{1}{m}\Sigma(n_s - mp_s)\,x_s$$

$$= \frac{1}{m}\Sigma(\delta n_s x_s)$$

As the expected value of δn_s is zero, the expected value of $\delta\bar{x}$ is zero, or the mean value found from repeated sampling of the mean of the sample is the same as the mean of the population.

Squaring both sides of the last equation above, we have

$$(\delta\bar{x})^2 = \frac{1}{m^2}\{\Sigma(\delta n_s^2 x_s^2) + 2\Sigma'(\delta n_s \delta n_t x_s x_t)\}$$

where Σ' indicates summation for all pairs of values of s and t for which s is not equal to t.

If we now assume a large number, N, of samples to have been taken and the corresponding values of $(\delta\bar{x})^2$ summed and the result divided by N, we obtain

$$\frac{1}{N}S(\delta\bar{x})^2 = \frac{1}{m^2}\left\{\Sigma\left(x_s^2 \frac{1}{N}S(\delta n_s^2)\right) + 2\Sigma'\left(x_s x_t \frac{1}{N}S(\delta n_s \delta n_t)\right)\right\}$$

where S denotes the summation in respect of the N samples. The left-hand side of this equation is the squared standard error of the mean of the sample, or $\sigma_{\bar{x}}^2$. On the right-hand side $\frac{1}{N}S(\delta n_s^2)$ is the $\sigma_{n_s}^2$ of equation (1) and $\frac{1}{N}S(\delta n_s \delta n_t)$ is given in equation (3). Hence

$$\sigma_{\bar{x}}^2 = \frac{1}{m^2}\{\Sigma[x_s^2 mp_s(1-p_s)] - 2\Sigma'(x_s x_t mp_s p_t)\}$$

$$= \frac{1}{m}\Sigma(p_s x_s^2) - \frac{1}{m}\{\Sigma(p_s x_s)\}^2$$

$$= \frac{1}{m}(\mu_2' - \bar{X}^2)$$

where μ_2' is the second moment, about the origin for x, of the distribution of the population. But $\mu_2' - \bar{X}^2 = \sigma_x^2$, therefore

$$\sigma_{\bar{x}} = \sigma_x / \sqrt{m} \qquad \ldots\ldots(5)$$

We thus find that the standard error of the mean is the ratio of the standard deviation in the population to the square root of the size of the sample.

10. This last result is of considerable use in statistical work. A large number of cases is recorded and the mean used to compare the particular experiment with another of a like kind. Is an actual difference between the means due to some cause other than random sampling? A practical application would be the comparison of the average profit from various classes of business for a number of years. The standard error of the profits in the various years would be obtained by taking the square root of the second moment about the mean and dividing it by the square root of the number of years; the quotient would give $\sigma_{\bar{x}}$ of (5). It is only by using the standard errors (or probable errors deduced from them) that we could say definitely whether a lower average profit in a certain part of the business was due to chance or to some causes requiring removal.

11. In § 5 of this chapter it was mentioned that we can often determine the actual curve to which the frequency distribution of a statistical measure tends. We saw, in Chapter IV, that β_1 and β_2 could, with the mean, be used to fix the frequency-curve if it is of the Pearson family of curves, and it follows that if we can find β_1 and β_2 for the frequency distribution of a statistical measure we shall have gone a long way towards fixing the form of the curve. If we write $\beta_1(x)$ and $\beta_2(x)$ as the moment ratios for the population distribution of x and $\beta_1(\bar{x})$ and $\beta_2(\bar{x})$ for the distribution of $\bar{x}$ (the sample mean) in repeated samples of size m, then it can be shown (see R. Henderson, *J. Inst. Actu.* XLI, 429) that

$$\left.\begin{array}{l} \beta_1(\bar{x}) = \beta_1(x)/m \\ \beta_2(\bar{x}) = 3 + \{\beta_2(x) - 3\}/m \end{array}\right\} \qquad \ldots\ldots(6)$$

Thus if the distribution of x is represented by the normal curve for which $\beta_1(x) = 0$ and $\beta_2(x) = 3$, it is seen that

$$\beta_1(\bar{x}) = 0 \text{ and } \beta_2(\bar{x}) = 3$$

and the distribution of $\bar{x}$ is also normal. Even if the distribution of x is not normal, it follows from equations (5) that $\beta_1(\bar{x})$ approximates to zero and $\beta_2(\bar{x})$ approximates to 3 if m is not too small.

12. The standard error of a standard deviation may be taken as $\dfrac{\sigma_x}{2}\sqrt{\left\{\dfrac{\beta_2(x)-1}{m}\right\}}$ for large samples and when the distribution of the population approximates to the normal curve of error (when $\beta_2(x) = 3$) the standard error becomes $\sigma_x/\sqrt{(2m)}$.

Another standard error which is often useful relates to the difference between two percentages or proportions. Thus if we make m_1 trials and the event happens n_1 times and in an independent m_2 trials we find n_2 happenings, in what circumstances can we conclude that $p_1 = p_2$ where the sample estimate of p_1 is n_1/m_1 and of p_2 is n_2/m_2? The solution might be useful when two rates of mortality, withdrawal or sickness are being compared.

If $p_1 = p_2 = p$, say, the standard error of the difference $n_1/m_1 - n_2/m_2$ is $\sqrt{\{p(1-p)(1/m_1 + 1/m_2)\}}$. We do not really know p, the underlying proportion to which the p_1 and p_2 of our experiments approximate, but on the hypothesis that there is a common value we may make an estimate of it from

$$(n_1 + n_2)/(m_1 + m_2).$$

This leads to a standard error of

$$\sqrt{\left\{\frac{n_1+n_2}{m_1+m_2}\left(1 - \frac{n_1+n_2}{m_1+m_2}\right)\left(\frac{1}{m_1}+\frac{1}{m_2}\right)\right\}}$$

As an example we may take (1) 1000 cases with 22 withdrawals giving a rate of withdrawal of ·0220 and (2) 600 cases with 19 withdrawals giving a rate of ·0317. Is the difference ·0097 significant? The combination of the two experiences gives 41/1600 or ·0256 as the rate of withdrawal. The standard error by the formula last given is ·0082. The difference is not significant. If however the numbers had all been three times greater, the standard error would have been ·005 and it would require little additional evidence to satisfy us that the difference is significant.

13. In similar ways it is possible to find the standard errors of the moments and constants, but this leads to the more theoretical parts of the subject with which it is inadvisable to deal in a book of this character. It is, however, necessary to call attention to the standard error of the coefficient of correlation owing to the importance of that function in statistical work.

As in the case of the mean, it will help to avoid confusion if we use a symbol, ρ, for the correlation coefficient in the population itself different from the symbol, r, for the coefficient calculated from a particular sample of m pairs of observations. From one sample to another r will vary about ρ and it has been shown that the standard error of r is, for large samples,*

$$\sigma_r = (1 - \rho^2)/\sqrt{(m-1)} \text{ approximately} \qquad \ldots\ldots(7)$$

If we do not know ρ we use r as an approximation to it.

14. This result was first given with $\sqrt{m}$ in the denominator by K. Pearson and L. N. G. Filon as an approximation when m is large (*Philos. Trans.* A, cxci, 231–41). Later R. A. Fisher (see *Biometrika*, x, 507–21) obtained the exact distribution for r when samples are drawn from a population following the normal correlation surface of p. 159 above. The closeness of the approximation by formula (7) as well as the form of the sampling distribution of r in such circumstances can be studied from tables given in *Tables for Statisticians*, Part II, Table xxxii or *Biometrika*, xi, 328 et seq. It can be seen from these tables that if $\rho = 0$, formula (7) gives a good value for σ_r even for very small values of m, but as ρ becomes larger the approximation is less satisfactory partly because the formula does not give a close value and partly because, even if σ_r be found closely, the

---

* When m is large $\sqrt{m}$ can be used for $\sqrt{(m-1)}$ here and in similar formulae.

distribution of r is such that + and − deviations are not equally likely and the usual rule that twice the standard error covers nearly the whole field may not apply. It is difficult to give a more definite statement but it may be of help to say that if $m > 400$ formula (7) can be used.* If however $m = 100$, care is needed in interpreting σ_r unless ρ is less than ·5; and if $m = 50$, unless ρ is less than ·3.

R. A. Fisher suggested (*Metron*, I, 1921) an useful transformation to

$$\tfrac{1}{2}\{\log_e (1+r) - \log_e (1-r)\}$$

which is distributed normally with a standard error of

$$1/\sqrt{(m-3)}$$

whatever the value of ρ.

15. As an application of formula (7) we may take the example in Chapter VII where we found that the coefficient of correlation between the age at maturity and the unexpired term of endowment assurances is ·254. It is not right however to assert that this coefficient exactly represents the correlation: the real measure may be greater or less, and considerations arise similar to those exemplified in §6. But there is another point in connection with a coefficient of correlation—we cannot even say that there is any real relationship till we have examined the standard error. In our example $m = 2870$ and $r = ·254$, so that $\sigma_r = \pm ·016$. In this case, therefore, the standard error is so small that the result is reliable, but if we had found $r = ·073$ with a standard error of ·05 it would have been impossible to say definitely that the correlation had not arisen merely from chance.

16. This brings us to an important application of the standard error in formula (7) which can be made safely even when m is as small as 30. If there is really no correlation, then $\rho = 0$ and the expression in (7) reduces to

$$\sigma_r = 1/\sqrt{(m-1)} \qquad \qquad \ldots\ldots(8).$$

---

* For $m = 400$, $\rho = ·9$, we find $\sigma_r = ·00957$, by formula (7) σ_r is ·00951; the distribution of r is described by mean $r = ·8998$, mode $= ·9011$, $\beta_1 = ·07402$, $\beta_2 = 3·1342$. This can only be roughly represented by a normal curve.

Thus, to go back to the example in Chapter VII, if we assume that there is no correlation, $r - \rho = \cdot 254$ with a standard error of $1/\sqrt{2869}$ or $\cdot 0187$. The difference, $r - \rho$, is well over twelve times the standard error; it is therefore almost impossible that the correlation was zero in the population from which the sample of 2870 may be supposed to have been drawn.

17. Formula (7) above is appropriate only for a coefficient of correlation calculated by the method described in Chapter VII. In using the fourfold table the standard errors are larger, as would be expected, because the grouping is rougher, and the formula by which they should strictly be calculated becomes complicated. The formula referred to gives as the standard error of r,

$$\frac{1}{\chi \sqrt{N}} \sqrt{\left\{ \frac{(a+d)(c+b)}{4N^2} + \psi_2^2 \frac{(a+c)(d+b)}{N^2} + \psi_1^2 \frac{(a+b)(d+c)}{N^2} \right.}$$
$$\left. + 2\psi_1\psi_2 \frac{ad-bc}{N^2} - \psi_2 \frac{ab-cd}{N^2} - \psi_1 \frac{ac-bd}{N^2} \right\}$$

where
$$\chi = \frac{1}{2\pi \sqrt{(1-r^2)}} e^{-(h^2+k^2-2rhk)/2(1-r^2)}$$

$$\psi_1 = \frac{1}{\sqrt{(2\pi)}} \int_0^{\frac{h-rk}{\sqrt{(1-r^2)}}} e^{-\frac{1}{2}x^2} dx$$

$$\psi_2 = \frac{1}{\sqrt{(2\pi)}} \int_0^{\frac{k-rh}{\sqrt{(1-r^2)}}} e^{-\frac{1}{2}x^2} dx$$

and it is assumed that the fourfold table is so arranged that $a+c > b+d$ and $a+b > c+d$, where a, b, c, and d have the meanings indicated on p. 173. The numerical work for finding the standard error of r for the example in Chapter IX is as follows

$$\psi_1 = \frac{1}{\sqrt{(2\pi)}} \int_0^{\frac{h-rk}{\sqrt{(1-r^2)}}} e^{-\frac{1}{2}x^2} dx = \frac{1}{\sqrt{(2\pi)}} \int_0^{\cdot 56821} e^{-\frac{1}{2}x^2} dx = \cdot 21505$$

by *Tables for Statisticians*, Part I, Table II

$$\psi_2 = \frac{1}{\sqrt{(2\pi)}} \int_0^{\frac{k-rh}{\sqrt{(1-r^2)}}} e^{-\frac{1}{2}x^2} dx = \frac{1}{\sqrt{(2\pi)}} \int_0^{\cdot 32230} e^{-\frac{1}{2}x^2} dx = \cdot 12639$$

(195)

13-2

$$\chi = \frac{1}{2\pi \sqrt{(1-r^2)}} e^{-(h^2+k^2+2hkr)/2(1-r^2)} = \frac{1}{2\pi \times \cdot 63900} e^{-\cdot 86744}$$

$$= \cdot 10462$$

$$\therefore \log \frac{1}{\chi} = \cdot 98039, \ \log \psi_1 = \overline{1} \cdot 33254 \text{ and } \log \psi_2 = \overline{1} \cdot 10171;$$

$\therefore$ the standard error of r is

$$\frac{1}{\cdot 10462 \sqrt{(4703)}} \sqrt{\{\cdot 02283 + \cdot 00145 + \cdot 00479}$$
$$+ \cdot 00252 - \cdot 00408 - \cdot 01015\} = \pm \cdot 018$$

18. The standard errors found by this method are larger than would result from formula (7) and in many cases are as much as three times as great—this actually happens in our example. The correct formula is rather troublesome, but *Tables for Statisticians*, Part I, Tables XXIII and XXIV, based on an approximation, minimise the arithmetical work. The approximation can be safely used except when the divisions of the correlation table differ extremely.

19. It will be noticed that, as we anticipated, all the expressions for the standard errors contain the square root of the number of cases in the denominator. We anticipated in the first paragraphs of this chapter that the standard error would decrease as the number of cases increased and we can now say that in each of the cases discussed the standard error varies inversely with the square root of the number of cases. The student should make it a rule to work out standard errors and he will find that much labour can be saved by using tables, usually of "probable errors", that have been published in the *Tables for Statisticians*.

The object in calculating standard errors is to prevent ourselves from reading too much from the means or other measures we have calculated, but we must not run to the opposite extreme and rely more on a standard error than the theory justifies. Thus, at certain points, our theory has assumed that the characteristics are distributed in a form approximating to a normal curve of error, and a good deal of evidence has been produced showing that this is a reasonable assumption for many characteristics when the number of cases exceeds 30, or for some characteristics with even smaller numbers. The assumptions imply that plus and minus errors are equally probable, but it would not be right to assert that the means of a sample of a J-shaped distribution are equally likely to fall above and below the true mean within twice the standard error, and formulae (6) above help to indicate this limitation.

20. We may now refer briefly to some practical points in "sampling". The essence of sampling is that we form an opinion of the whole by examining a sample of it, and error may arise (1) owing to bias in making up the sample or (2) owing to the particular sample giving a wide deviation from the whole because it is based on a small number of cases.

It is, usually, not difficult to guard against bias in actuarial or sociological practice. For instance, if we require to estimate the mortality of lives assured we might collect information merely in respect of persons whose names begin with A. This would give fewer cases, but there is no reason to suspect that such lives differ from those whose names begin with the other letters of the alphabet. The selection of a particular letter might, however, lead to suspicion if it could introduce a question of race in a mixed community, e.g. in Alsace-Lorraine; if we worked with people whose names begin with W we should exclude those of French extraction but include those of German extraction. An alternative is to take one case in, say, each hundred, e.g. the mortality of lives assured could be investigated by examining from the registers of the insurance offices every hundredth case.

Sampling of this kind is useful in social investigations where we may, perhaps, want to examine the home conditions of school children and cannot hope to get from every home particulars of the health, occupation or habits of the residents. We might, however, be able to make an exhaustive examination of 2,000 or 3,000 cases. With a free hand it is not difficult to obtain a random sample, and a little thought and common sense is all that is required.

The other risk of error lies in the fact that we have only a small sample, and it is here that the subject is connected with that of "standard errors". If we may assume that the sample is chosen at random and, though not of itself small, is small compared with the population from which it is drawn, then we can follow the methods indicated in the earlier part of this chapter.

21. Special circumstances, however, arise in some experiments, and one type of case may be specially mentioned. It is frequently necessary to test the comparative yields of different varieties of the same plant. The trouble in such a case is that plots placed far apart even in a small field produce widely different results, but small adjacent plots resemble each other. In order to make a fair comparison we ought, therefore, to have a number of pairs of adjacent plots. The comparison is made between a number of pairs and we are concerned with the differences between these pairs and must work out

$$\sigma^2 = \frac{S(y-x)^2}{m}$$

where m is the number of "pairs", and x and y are the corresponding members of a pair measured from their means.

It is important to distinguish this sort of case, where the pair formed from adjacent plots is the unit, from the different case where we draw a sample of m_1 observations from one record with a standard deviation of σ_1 and a second sample of m_2 from another record in which the standard deviation is σ_2. In this case the standard deviation of the difference between the two means is given by

$$\sigma^2 = \frac{\sigma_1^2}{m_1} + \frac{\sigma_2^2}{m_2}$$

This assumes that there is no correlation between the variables, but in the "pairs" problem we have arranged "pairs" because we expect correlation. Algebraically the correlation is indicated by the xy term of $S(y-x)^2 = S(y^2 - 2xy + x^2)$.

(198)

It will be appreciated from what has been written elsewhere in this chapter that it is assumed that the samples are sufficiently large to justify the assumption that the σ's calculated from the samples can be treated as the standard deviations of the population.

The use of the wrong formula may lead to erroneous conclusions: the actual difference between the means may be 30, the standard error by $\sigma^2 = \dfrac{S(y-x)^2}{m}$ may be 6, and by $\sigma^2 = \dfrac{\sigma_1^2}{m_1} + \dfrac{\sigma_2^2}{m_2}$ may be 12. Judged by the former the difference is almost certainly significant; judged by the latter it is doubtful.

The kind of problem indicated might arise whenever it is necessary to compare the results of alternative methods in changing conditions, and the theory which was worked out primarily to test yields may prove valuable elsewhere.

CHAPTER XI

THE TEST OF GOODNESS OF FIT

1. When the values of ordinates and areas were calculated in the examples of the various types of frequency-curves, no systematic attempt was made to test the graduations in order to ascertain whether the results obtained were reasonable. Actuaries have generally been in the habit of imposing on the graduated values of any table on which they may have been working, rough checks which have amounted to a comparison of the totals in various groups and an inspection of the changes of sign in the differences between the graduated and ungraduated figures. The problem of the goodness of fit needs, however, more accurate treatment; for inspection, even when aided by the calculation of a standard error for each group, can only tell that certain differences are large, and if the standard error be exceeded in two or three cases, it is impossible to say whether the excesses are in any way balanced by equalities in the rest of the graduation. A test is required which will give some measure of the disagreement as judged by the whole graduation.

2. Now, if there be N observations distributed in n groups, the numbers in the group being m_1', m_2', ..., m_n', we have to find a criterion to enable us to decide when the series m_1, m_2, ..., m_n will be a legitimate graduation. We may clearly take a legitimate graduation to be one in which the observed values (m') do not differ from the theoretical (m) by more than the deviations that would be expected in random sampling. What we require to know is not the probability that the particular series of m''s will occur if the m's represent the theory, but the probability that the m''s, or an *equally likely or less likely* series, will arise. To appreciate the difficulties of the problem we may consider the simplest case, that of a coin-tossing experiment, and suppose that a coin has been tossed six times and come down 4 heads and 2 tails. The "graduation" we make is 3 heads and 3 tails, and to test it we require to find the probability of obtaining a result as unlikely, or more unlikely than the observed one. This probability is the same as that of getting any one of the following results:

6 heads and		0 tails
5	,,	1 ,,
4	,,	2 ,,
2	,,	4 ,,
1	,,	5 ,,
0	,,	6 ,,

It is impossible to calculate such probabilities directly, even when the simple probabilities leading to the deviations are known, in any but the easiest cases; but when we do not know the simple probabilities, or the case is a complicated one, a further difficulty is introduced owing to our inability to tell from *a priori* reasoning which of the possible cases are more or less likely than that which has actually arisen. It would, for instance, be difficult to say, without a large amount of arithmetical work, when 20 dice were being thrown, whether the probability of getting ten "sixes" or more was greater than that of getting two "sixes" or fewer; but this is an extremely simple case compared with the general proposition in which deviations over a series of numbers have to be considered.

3. If it is assumed in any measurement on one subject that the deviations from the mean take the form of the "normal curve of error", and it is required to estimate the chance of obtaining deviations greater than a certain value (t, say), it will be necessary to sum all values of the normal curve beyond t on each side of the mean, i.e. we must take

$$\int_{-\infty}^{-t} e^{-\frac{1}{2}x^2} dx + \int_{t}^{\infty} e^{-\frac{1}{2}x^2} dx = 2 \int_{t}^{\infty} e^{-\frac{1}{2}x^2} dx$$

and divide the result by the area of the whole curve, i.e. by the total deviations. Assuming that there are two measurements instead of one (the exposed to risk, for instance, at two ages), the deviations are as it were, in two directions instead of one, and it is necessary to take an expression with two variables instead of one. The expression analogous to the normal curve is the correlation surface

$$z = z_0 e^{-\frac{1}{2}\left\{\frac{x^2}{\sigma_1{}^2} - \frac{2xyr}{\sigma_1\sigma_2} + \frac{y^2}{\sigma_2{}^2}\right\}\big/(1-r^2)}$$

with which we have already dealt. The integrations must be performed for both variables from t and t' onwards, and compared with the total. If there are n measurements it becomes necessary to deal with a function of n variables, and this will give the reader a slight idea of the problem from the mathematical point of view, and suggest that he will expect the quotient of two n-fold integrals to give the probability. The next step is to reduce these n-fold integrals to the form of ordinary integrals, and it has been shown* that the result

$$P = \frac{\displaystyle\int_\chi^\infty e^{-\frac{1}{2}x^2} x^{n-1} dx}{\displaystyle\int_0^\infty e^{-\frac{1}{2}x^2} x^{n-1} dx} \; \dagger$$

is reached. In this expression χ stands for a complex function depending on the n variables from which the expression was evolved, and measures the position that is indicated by the probability of the particular distribution, the test for the graduation of which is required.

---

* Originally by K. Pearson: "On the criterion that a given system of deviations from the probable in the case of a correlated system of variables is such that it can be reasonably supposed to have arisen from Random Sampling," *Phil. Mag.*, July 1900. A short proof has been given by H. E. Soper in "Frequency Arrays".

† A table of P for all values of $n' = n + 1$ from 3 to 30, corresponding to χ^2 from 1 to 30, with a few additional values and auxiliary tables for the calculation of further values, is given in *Tables for Statisticians*, Part I. An abridged table is given in Appendix IX.

4. Before a measure of the probability P can be obtained a value for χ must be found from the statistics of the particular graduation, and in the paper to which reference has already been made its value is shown to be such that

$$\chi^2 = S \left\{ \frac{(m_r - m_r')^2}{m_r} \right\}$$

It is natural, almost necessary, to use the square of the difference in order that negative differences may, equally with positive differences, increase the improbability of the system, while a ratio is required to bring into account the size of the group, for an error of 15 in a group of 20 would be very large, but in a group of 1,000 would be negligible.

5. The practical aspects of the test of goodness of fit and its application may now be dealt with.

(1) If the facts representing the graduated and ungraduated figures are only available in groups, then the value of the probability by the test will, as a rule, be lower as the number of groups is increased. This practical point should be borne in mind as it sometimes happens that graduations are tested in groups of, say, 5 years of age; but the graduated figures for individual ages are then used unreservedly, though, strictly speaking, they may be no better than interpolated values.

(2) The test assumes a distribution, and would not be applicable if the numbers were a series of ordinates, though the application of the test would probably give a fair idea of the goodness of fit if a large number of ordinates had been given in the series.

(3) The tails of the experience will be very small and never fit exactly. We ought to take our final theoretical groups to cover as much of the tail area as amounts to at least a unit of frequency in such cases.

(4) If the number of observations be multiplied by t, say, and the deviations are also multiplied by t, then the value of χ^2 will be multiplied by the same figure, and the test will show that the fit is worse. This may seem strange at first, but a

little consideration will show that it is reasonable. As a large number of cases will give smoother series than a small number, it follows that if two results are proportionally the same in two examples having the same theoretical distribution but different total frequencies, the one with greater frequency is less probable than the one with less frequency. The probability of a result as bad as, or worse than, three heads and one tail in coin-tossing (two heads and two tails being the theoretical result) is ·625; but the probability of a result as bad as, or worse than, $3 \times 2 = 6$ heads and $1 \times 2 = 2$ tails is ·289. It follows that if a distribution is based on, say, 103,480 cases and the figures are reduced to a total of 1,000 to show the distribution of the cases, then a graduation tested as if 1,000 were the total frequency will give the impression that the graduation is far closer than it really is.

(5) I have found, in applying the test, that when the numbers dealt with are very large the probability is often small, even though the curve appears to fit the statistics very closely. The explanation may be that the statistics with which we deal in practice nearly always contain a certain amount of extraneous matter, and the heterogeneity is concealed in a small experience by the roughness of the data. The increase in the number of cases observed removes the roughness, but the heterogeneity remains. The meaning, from the curve-fitting point of view, is that the experience is really made up of more than one frequency-curve; but a certain curve, approximating to the one calculated, predominates. Another possible explanation is that our solution of the problem depends on the assumption of a mathematical expression which does not give exactly the distribution of deviations and when we deal with a large experience the approximate nature of the assumptions is revealed.

(6) What is the actual value of P at which a good fit ends and a bad one begins? It is impossible to fix such a value. We have merely a measure of probability for the whole table, and if the odds against the graduation are twenty or thirty to one

the result is unsatisfactory; if they are ten to one the graduation is not unreasonable, but the exact value when a result
must be discarded cannot be given. As, however, it is clearly
impossible to imagine any test which can fix an absolutely
definite standard, there is no reason for objecting to the
particular method because it fails to do so.

(7) It is sometimes thought that the introduction of
additional constants must necessarily improve the fit of a
curve. It may do so in some cases, but it is quite possible to
take a curve with ten constants and find it gives a worse result
than another having only three. Besides this, there is the
possibility of undergraduation; we must not expect to reach
a very high value for P, e.g. ·95. If we make an experiment in
coin-tossing, it is unlikely that a single experiment will give
a distribution very close to the theoretical. If therefore we are
estimating the probability of getting that result or worse, we
shall only rarely get a very high or a very small value for that
probability. We shall do so occasionally, but we must not
expect it and it is wise to look for explanations when any
graduation gives a very high or very low value of P.

(8) It may sometimes be advisable to use a curve giving a
worse agreement than another for simplicity, or for reasons
such as those which prompt actuaries to employ Makeham's
hypothesis.

6. In a paper "On the Comparative Reserves of Life
Assurance Companies, etc." (*J. Inst. Actu.* XXXVII, 458–9),
George King remarked that it is permissible to use the H^M
Model Office for the O^M; and it will be interesting to apply the
formulae given above to see what is the probability of the O^M
distribution if the H^M be taken as the theoretical distribution.

In the table on p. 206 there are ten groups, and $\chi^2 = 1\cdot79$,
and *Tables for Statisticians* give $P = \cdot999438$ and $\cdot991468$ when
$\chi^2 = 1$ and 2 respectively. It is not, however, sufficient to test
for 100 new policies. 950 would reduce the probability to
about ·05, which means that in only one case out of twenty
would a random sampling lead to a system of deviations from

the H^M as great as that shown by the O^M. This result will remind the student of the great danger of dealing with percentages without considering the actual number of cases investigated. King's other table, which is of greater importance in his work (policies according to attained age), shows a much closer agreement, as $P = \cdot 831051$ for 10,000 cases.

Central age in group	Policies issued arranged in age-groups		$O^M - H^M$		(Square of $O^M - H^M$)/H^M
	H^M	O^M	+	−	
20	6·97	7·30	·33	...	·02
25	17·75	20·45	2·70	...	·41
30	21·04	23·11	2·07	...	·20
35	18·41	18·40	...	·01	·00
40	13·82	13·05	...	·77	·04
45	9·45	8·44	...	1·01	·11
50	6·23	5·07	...	1·16	·22
55	3·51	2·58	...	·93	·25
60	1·97	1·20	...	·77	·30
65	·85	·40	...	·45	·24
...	100·00	100·00	5·10	5·10	$\chi^2 = 1 \cdot 79$

7. We will now revert to § 2 of this chapter where in stating the problem it was said that the N observations were distributed in $n + 1$ groups. As we have only N observations to distribute we can only choose n groups, for having fixed those n groups the last one is necessarily fixed; freedom of choice is restricted to this extent, and in any problem where the method is used the number of groups where freedom of choice is possible must be borne in mind. This is implied in the proofs leading up to the formulae which have been given. Now, following on this argument the reader may ask whether it is fair in comparing a Type I and a Type III graduation of certain material to use the same value of n when there are four constants necessary to reach the former and three to reach the latter. He may ask "are we not really restricting our freedom of choice more in the former case than the latter because, to take an extreme case, we should reproduce a distribution of only four groups exactly with Type I and alter it, that is have freedom of choice, if we use Type III?"

8. Before we deal with this question we may explain that in the previous paragraphs of this chapter two distinct problems have been covered by the one word "graduation". These problems are

I. Given a theoretical distribution, to ascertain the probability of getting an actual distribution or an equally likely or a less likely one.

Here is an example which compares the theoretical number of "heads", when six coins are tossed, with an actual distribution.

No. of "heads" (1)	Theoretical (2)	Actual (3)	(2) − (3) (4)	Square of (4) (5)	(5)/(2) (6)
0	1	0	1	1	1·00
1	6	6	0	0	·00
2	15	12	3	9	·60
3	20	23	− 3	9	·45
4	15	18	− 3	9	·60
5	6	3	3	9	1·50
6	1	2	− 1	1	1·00
Total	64	64			$\chi^2 = 5\cdot15$

If $n' = 7$ and $\chi^2 = 5\cdot15$, then $P = \cdot52$.

II. Given a graduation of an actual distribution, to ascertain the probability that the deviations will be the same as or greater than those found.

The answer depends on the number of constants in the formula used for graduation. If there are r constants we should deduct r from the number of groups instead of deducting unity as is, in effect, done in the last example for $n' = n - 1$. The mean is used to fix the position of the curve and must be counted as a constant. Consequently we must deduct 3 if the normal curve is used (i.e. one for the total number of cases, one for the mean and one for the s.d.), 4 for Type III and 5 for the main types.* Generally speaking, the same result is obtained if the number of moments used in the calculations be deducted. This

* In *Tables for Statisticians*, Part I, $n' = n - 1$, and one is, therefore, already deducted. It follows that $n' - 2$ would be the number to be used if the normal curve has been used for graduating.

(207)

gives the theoretical answer to the question raised at the end of §7 above. It is not always easy to interpret the number of moments in applying the rule: thus we choose between Type III and Type V by using the fourth moment, though there are only three moments needed to find the constants. Again if in Type I the start of the curve is fixed, three moments only are used; while if the range is fixed, only two moments are used (and in effect the number of constants is similarly decreased). If we make a rough attempt at a graduation by a Type I curve using four unadjusted moments and then vary the start of the curve as indicated in Chapter V, §10, then the final graduation only uses three moments. It can be argued that the full number of constants has been assumed and four moments have really been used.

9. The example given for Problem I in §8 can be used to explain the point mentioned in §5 about undergraduation. We may, on a particular occasion, reach an actual distribution identical with the theoretical. χ^2 will then be zero and P will be unity. Similarly we may reach a distribution so far from the theoretical as to seem well-nigh impossible. One of these exceptional cases may appear and if we repeat the experiment long enough we shall get distributions giving all values of P. Similarly with graduation; we are unlikely, if we know the right form of curve, to find a value of P that is infinitesimally small or very near to unity, but neither is impossible.

10. When we merely want to compare several graduations of the same distribution we can often stop our work after the calculation of χ^2. Thus if we make graduations by Type I using various adjustments or compare them with Type A or Type B using the same number of constants, the lowest value of χ^2 shows the closest graduation. Even if the number of constants differs, the value of χ^2 shows which graduation is actually closest and for some actuarial work this may be more important than the study of the probabilities.

Bearing in mind that there are difficulties in interpreting the number of degrees of freedom in some cases, we may consider what is implied when we use the solution of Problem I for Problem II. All the old applications of the (P, χ^2) test were made in this way. In such circumstances we are saying, in effect, that the graduation is a theoretical distribution not necessarily obtained from the actual distribution but by general reasoning or from other previous experience, and that we are measuring the probability of divergences from that theoretical distribution as great as or greater than those of the actual distribution.

The points set out in these paragraphs are mentioned because it is well to be reminded that we must not read into a good general test of graduation a refinement which is neither justified by the underlying theory nor required in practical work.

11. Reference may here be made to a test of a graduation of a mortality table. The data are expressed as "exposed to risk" (E_x) at each age (or group of ages) and "deaths" (θ_x). A graduation of the rates of mortality is made and the "expected deaths" (θ'_x) are calculated by multiplying the values of E_x by the appropriate graduated rates of mortality (q_x). We have, therefore, graduated the series

$$\theta_x, \; E_x - \theta_x; \; \theta_{x+1}, \; E_{x+1} - \theta_{x+1}; \; \text{etc.}$$

by
$$\theta'_x, \; E_x - \theta'_x; \; \theta'_{x+1}, \; E_{x+1} - \theta'_{x+1}; \; \text{etc.}$$

The E_x is fixed in each pair, so, if there are 40 ages, there are only 40 degrees of freedom, not 80. But the χ^2 should be calculated from all the 80 values, although when E_x is large relatively to θ, as it is at nearly every age, the $E - \theta$ terms give zero elements. It will be easier for the reader to follow this argument if he bears in mind that the total of the θ's need not be reproduced exactly by the θ''s. Deduction will have to be made from the 40 degrees of freedom for the number of constants used in the graduation.

CHAPTER XII

THE CORRELATION RATIO—CONTINGENCY

1. We have seen that we can reasonably use the coefficient of correlation when regression is linear, that is when the means of the columns (and the means of the rows) are approximately in a straight line; but in other circumstances its use is open to objection. In the present chapter other methods are described which are not open to the same objection. We shall deal first with a function known as the "correlation ratio" (η), which is a useful measure in some cases.

The value of η_{yx} is given by

$$\eta_{yx}^2 = \frac{S\{n_x(\bar{y}_x - \bar{y})^2\}}{N\sigma_y^2} \qquad \ldots\ldots(1)$$

where $\bar{y}_x$ is the mean of the y's corresponding to the particular array x, n_x is the number of cases in the array x, N the total frequency, σ_y the standard deviation of the y's and $\bar{y}$ is the mean of all the y's. The summation extends over all the arrays. In a similar way we can work from the y-arrays and have

$$\eta_{xy}^2 = \frac{S\{n_y(\bar{x}_y - \bar{x})^2\}}{N\sigma_x^2} \qquad \ldots\ldots(2)$$

These values of η will not be the same except in the limiting case when regression in both directions is linear and then $\eta_{yx} = \eta_{xy} = r$. It will be seen that the correlation ratio η_{yx} can alternatively be expressed as the ratio of the standard deviation of the means of the y-arrays, each array being weighted with the number in it, to the standard deviation of the y's.

Taking the example on p. 142 we should find η as follows:

Mean unexpired term in each column $\bar{y}_x$	Deduct mean of whole (20·312) $\bar{y}_x - \bar{y}$	$(\bar{y}_x - \bar{y})^2$	n_x	$n_x(\bar{y}_x - \bar{y})^2$
10·333	− 9·979	99·6	6	598
13·250	7·062	49·9	4	200
13·176	7·136	50·9	17	865
16·113	4·199	17·6	62	1,091
17·230	3·082	9·50	584	5,548
20·141	·171	·029	643	19
21·877	+ 1·565	2·45	1,098	2,690
21·665	1·353	1·83	388	710
21·500	1·188	1·41	60	85
27·625	7·313	53·5	8	428
...	...	...	2,870	12,234

$$\therefore \qquad \eta^2_{yx} = \frac{12234}{2870 \times (7 \cdot 6067)^2} = \cdot 07367$$

or $\qquad \eta_{yx} = \cdot 2708$

The figure 7·6067 is the value of σ_2 on p. 149, multiplied by 5 the unit of grouping.

Working similarly with the maturity ages, we obtain the following:

$\bar{x}_y - \bar{x}$	$(\bar{x}_y - \bar{x})^2$	$n_x(\bar{x}_y - \bar{x})^2$
− 3·48	12·11	678
2·20	4·84	832
1·38	1·90	821
·64	·41	273
+ ·35	·12	81
·65	·42	226
2·71	7·34	1,813
3·81	14·52	1,118
5·27	27·77	222
7·77	60·37	60
...	...	6,124

$$\therefore \qquad \eta^2_{xy} = \frac{6124}{2870 \times (5 \cdot 6818)^2} = \cdot 06610$$

or $\qquad \eta_{xy} = \cdot 2571.$

The arithmetical processes described in §§ 11–13 of Ch. VII supply us with most of the figures required.

(211)

2. We may now go back to formula (1) and rearrange the denominator. Remembering that the square of the standard deviation can be found by squaring the difference between each observation, o_{xy}, and the mean (see Chapter III, § 14), we have

$$N\sigma_y^2 = \underset{x\,y}{SS}(o_{xy} - \bar{y}_x + \bar{y}_x - \bar{y})^2$$
$$= \underset{x\,y}{SS}(o_{xy} - \bar{y}_x)^2 + \underset{x}{S}\{n_x(\bar{y}_x - \bar{y})^2\} + 2\underset{x}{S}\{(\bar{y}_x - \bar{y})\underset{y}{S}(o_{xy} - \bar{y}_x)\}$$
$$= \underset{x\,y}{SS}(o_{xy} - \bar{y}_x)^2 + \underset{x}{S}n_x(\bar{y}_x - \bar{y})^2$$

as the final expression in the previous line vanishes.

Consequently

$$\eta_{yx}^2 = \frac{\underset{x}{S}\{n_x(\bar{y}_x - \bar{y})^2\}}{\underset{x}{S}\{n_x(\bar{y}_x - \bar{y})^2\} + \underset{x\,y}{SS}(o_{xy} - \bar{y}_x)^2} \qquad \ldots\ldots(3)$$

$\underset{x}{S}\{n_x(\bar{y}_x - \bar{y})^2\}$ measures the amount of variation between arrays, while $\underset{x\,y}{SS}(o_{xy} - \bar{y}_x)^2$ measures the amount of variation within the arrays. Neither part of the denominator can be negative; therefore

$$1 \geqslant \eta_{yx}^2 \geqslant 0$$

It also follows from (3) that for η^2 to be large $\underset{x}{S}\{n_x(\bar{y}_x - \bar{y})^2\}$ must be large as compared with $\underset{x\,y}{SS}(o_{xy} - \bar{y}_x)^2$; in other words, the larger η^2 becomes, the greater the variation in the means of the arrays compared to the variations within the arrays. Also the smaller η^2 becomes, the less important are the differences in the means of the arrays.

3. The correlation ratio may be used for three main purposes:

(*a*) to measure the relationship between x and y—this has already been shown in the example,

(*b*) to test whether there is any real difference in the array means, $\bar{y}_x$, other than what might be expected from sampling,

(*c*) to test whether it is reasonable to regard the regression line as a straight line.

In dealing with (b) and (c) we must suppose that the distribution of y for each x array is not far from a "normal" distribution and that the standard deviations of y arrays for given x are approximately equal. Under these conditions it may be shown that, for the test mentioned in (b) above, if the array means, say k in number, in the population are all equal, so that the population value of η_{yx}^2 is zero, then in a sample of N pairs of values of x and y,

(i) the expected value of η^2, say,
$$\overline{\eta^2} = (k-1)/(N-1) \qquad \ldots\ldots(4)$$

(ii) the standard error
$$\sigma_{\eta^2} = \frac{1}{N-1}\sqrt{\{2(k-1)(N-k)/(N+1)\}} \qquad \ldots\ldots(5)$$

Unless, therefore, the observed η^2 is larger than, say, $\overline{\eta^2} + 2\sigma_{\eta^2}$, we cannot feel confident that it is significant, or that the means of the arrays in the population differ. The distribution of η^2 is however very skew if the number of arrays is small, so that a deviation of twice the standard error has to be viewed as indicated in Chapter X, § 19.

Under the same conditions we can show that, for (c) above, if in the population the means of the arrays $(\bar{y}_x)$ lie on a straight line, i.e. regression is linear, and $\eta^2 - r^2 = 0$, then in a sample of N pairs of values of x and y the ratio
$$(\eta^2 - r^2)/(1 - r^2)$$
will have

(i) an expected value of
$$(k-2)/(N-2) \qquad \ldots\ldots(6)$$

(ii) a standard error of
$$\frac{1}{N-2}\sqrt{\{2(k-2)(N-k)/N\}} \qquad \ldots\ldots(7)$$

and we can then judge of the departure from linearity of regression in the sample by applying a similar test to that in (b).

4. In the same numerical example (see the first table in § 1), $k = 10$, $N = 2870$ and the values from formulae (4) and (5) are
$$\overline{\eta^2} = \cdot0031, \qquad \sigma_{\eta^2} = \cdot0015$$

We have actually $\eta^2 = \cdot 0737$ so that there is a real difference in the array means.

If we take the second table and use the test of formulae (6) and (7), we find $\eta^2 = \cdot 06610$ so near to $r^2 = \cdot 06474$ that the ratio $(\eta^2 - r^2)/(1 - r^2)$ is $\cdot 0014$. The expected value is $\cdot 0028$ and the standard error $\cdot 0014$. This table shows linearity. The first table would hardly have done so.

5. We may now turn to the theory of contingency which gives us another way of approaching correlation and can be used when the regression is not linear or when the facts are given in a non-quantitative form with a greater number of divisions than those of the fourfold tables discussed in Chapter IX. The principle underlying the theory of contingency is that a comparison is made between the given table and a corresponding table having the same marginal totals but with no correlation. The first step, therefore, is to see how to make a table without correlation, and a little consideration will show that all we have to do is to split up the total of any column in proportion to the distribution of entries in the final total column. Thus, the first column would be

| Unexpired Term | ... | ... | 2 | 7 | ... |
| Frequency with no correlation | | $6 \times \frac{56}{2870}$ | $6 \times \frac{172}{2870}$ ... |

and the remaining part of the table would be formed in a similar way. Now as each column is formed in proportion to the total, the mean of each column must be the same as the mean of the total, which shows at once from the definition that no correlation can exist in such a table.

6. The following table shows the figures exhibiting no correlation in ordinary type, and those actually occurring in small type. Now, if these two sets of figures coincide exactly in any particular case, there is clearly no correlation in the table; if they differ slightly there is a slight amount, and if they differ greatly there is a considerable amount of correlation, and we come therefore to the conclusion that an alternative method of finding the correlation between two things is by measuring

Central unexpired term of Endowment Assurances	CENTRAL AGES AT MATURITY										Total
	30	35	40	45	50	55	60	65	70	75	
2	·1	·1	·3	1·2	11·4	12·5	21·4	7·6	1·2	·2	56
	2	...	...	2	26	6	14	6	...	...	
7	·4	·2	1·0	3·7	35·0	38·6	65·8	23·2	3·6	·5	172
	1	1	2	6	62	36	40	22	2	...	
12	·9	·6	2·6	9·3	87·8	96·8	165·4	58·4	9·0	1·2	432
	...	2	9	17	117	99	127	52	8	1	
17	1·4	·9	3·9	14·4	135·3	149·0	254·4	89·9	13·9	1·9	665
	3	...	6	24	145	155	237	84	11	...	
22	1·4	·9	4·0	14·6	137·2	151·0	257·8	91·1	14·1	1·9	674
	...	1	...	3	133	167	271	78	20	1	
27	1·1	·8	3·2	11·6	109·5	120·6	205·9	72·7	11·2	1·4	538
	...	...	...	9	90	123	231	71	11	3	
32	·5	·4	1·5	5·3	50·3	55·3	94·4	33·4	5·2	·7	247
	...	...	...	1	11	49	127	49	8	2	
37	·2	·1	·5	1·7	15·7	17·2	29·4	10·4	1·6	·2	77
	...	...	...	...	...	6	49	22	...	...	
42	·0	·0	·0	·2	1·6	1·8	3·1	1·1	·2	·0	8
	...	...	...	...	...	2	2	3	...	1	
47	·0	·0	·0	·0	·2	·2	·4	·2	·0	·0	1
	...	...	...	...	...	...	...	1	...	...	
Total	6	4	17	62	584	643	1,098	388	60	8	2,870

the difference between the figures in the actual correlation table and those that would have arisen if there had not been any correlation. In Chapter XI we discussed a method of measuring the goodness of fit (or amount of agreement) between two sets of figures, and this suggests that we might calculate χ^2 by squaring the difference between each pair of figures in the table and dividing the result by the frequency when there is no correlation. The reason for choosing the figure from the table with no correlation as the divisor is that it always has a value, while the correlation table may give a frequency of zero.

7. As it is clear that χ^2 will give a measure of the association, it will be interesting to see the connection between it and the coefficient of correlation r; and the following proof shows that if the correlation table can be approximately represented by the normal correlation surface, then where the number of groupings is large

$$r = \sqrt{\frac{\phi^2}{1+\phi^2}}$$

where $$\phi^2 = \chi^2/N \qquad \qquad ......(8)$$

Using the same notation as that of Chapter VIII, the frequency with no correlation is given by

$$Z'_0 = \frac{N}{2\pi\,\sigma_1\sigma_2}\,e^{-\frac{1}{2}\left(\frac{r^2}{\sigma_1{}^2}+\frac{y^2}{\sigma_2{}^2}\right)}$$

while that with correlation is

$$Z = \frac{N}{2\pi\,\sqrt{(1-r^2)}\,\sigma_1\sigma_2}\,e^{-\frac{1}{2}\frac{1}{1-r^2}\left(\frac{r^2}{\sigma_1{}^2}-\frac{2rxy}{\sigma_1\sigma_2}+\frac{y^2}{\sigma_2{}^2}\right)}$$

Then $\phi^2 = \displaystyle\int_{-\infty}^{+\infty}\int_{-\infty}^{+\infty}\frac{(Z-Z'_0)^2}{N Z'_0}\,dx\,dy$

$$= \frac{1}{N}\int_{-\infty}^{+\infty}\int_{-\infty}^{+\infty}\left(\frac{Z^2}{Z'_0}-2Z+Z'_0\right)dx\,dy$$

$$= \frac{1}{2\pi}\left\{\frac{1}{1-r^2}\int_{-\infty}^{+\infty}\int_{-\infty}^{+\infty}e^{-\frac{1}{2}\left\{r'^2\frac{1+r^2}{1-r^2}-\frac{4rx'y'}{1-r^2}+y'^2\frac{1+r^2}{1-r^2}\right\}}dx'\,dy'\right.$$

$$-\frac{2}{\sqrt{(1-r^2)}}\int_{-\infty}^{+\infty}\int_{-\infty}^{+\infty}e^{-\frac{1}{2}\left\{r'^2\frac{1}{1-r^2}-\frac{2rx'y'}{1-r^2}+y'^2\frac{1}{1-r^2}\right\}}dx'\,dy'$$

$$\left.+\int_{-\infty}^{+\infty}\int_{-\infty}^{+\infty}e^{-\frac{1}{2}(x'^2+y'^2)}dx'\,dy'\right\}$$

where $x' = x/\sigma_1$ and $y' = y/\sigma_2$

$$= \frac{1}{1-r^2}\,\frac{1}{\sqrt{\left\{\left(\frac{1+r^2}{1-r^2}\right)^2-\frac{4r^2}{(1-r^2)^2}\right\}}}$$

$$-\frac{2}{\sqrt{(1-r^2)}}\,\frac{1}{\sqrt{\left\{\frac{1}{(1-r^2)^2}-\frac{r^2}{(1-r^2)^2}\right\}}}+1$$

by (vi) of Appendix IV

$$= \frac{1}{1-r^2}-2+1$$

$$= \frac{r^2}{1-r^2}$$

or $\qquad r = \pm\sqrt{\dfrac{\phi^2}{1+\phi^2}}$

8. The result just obtained may be considered a little more closely.

(1) It shows that r must lie between -1 and $+1$.

(2) As the value of ϕ^2 will not be affected by the order of the columns (or rows), it is permissible to interchange them, provided, of course, the whole column (or row) be moved at once.

(3) The proof shows that r will not necessarily be obtained exactly if a very small number of groups is used, because by using the integral calculus an infinite number of groups was assumed.

(4) We also assumed, however, that we were dealing with smooth series; but as χ^2 is a measure of the goodness of fit between the correlation and no-correlation figures, a large number of groups gives undue prominence to the chance deviations due to the use of a random sample, and the value of r found from that of ϕ^2 may differ considerably from the value reached by the xy-moment. Too fine a grouping may give a less accurate result than a less fine one.

9. These conclusions are borne out by practical work, and any student who cares to go into the subject can find the value of r by the two methods from a large table, using various groupings, and he will see that the best agreements are obtained when the grouping is neither very fine nor very rough. But this general remark indicates a difficulty, for the student will naturally wonder how he is to group his figures in order to reduce them to a suitable number of classes. If he is dealing with facts distributed according to age, he can take groups of ten years instead of the finer grouping of five or three years or he may lump together the small groups at the ends. He will find that equal frequencies give better results than equal ranges when the material is divided into six (or less) classes, but when there are more than six classes equal ranges should be taken. This rule can only be applied broadly: we cannot from the nature of the data make exactly equal groups of our frequencies but must be content with something approaching equality. In order to indicate how we may proceed and how the numerical work is done, the following table has been prepared from that of p. 215.

Central unexpired term	CENTRAL AGES AT MATURITY				Total
	50 and under	55	60	65 and over	
2, 7, 12	154·6 (247)	147·9 (141)	252·6 (181)	104·9 (91)	660
17	155·9 (178)	149·0 (155)	254·4 (237)	105·7 (95)	665
22	158·1 (137)	151·0 (167)	257·8 (271)	107·1 (99)	674
27	126·2 (99)	120·6 (123)	205·9 (231)	85·3 (85)	538
32 and over	78·2 (12)	74·5 (57)	127·3 (178)	53·0 (86)	333
Total	673	643	1,098	456	2,870

10. The totals are not all equal to one another: the 1,098 cases maturing at age 60 prevent this, but they are far more nearly equal than the totals in the original table. We now work out χ^2 and find that its value is 198·8.* Hence

$$\phi^2 = \frac{\chi^2}{N} = ·0693$$

and the coefficient of contingency is ·254. This differs from the figures given for η in §1† and both may differ from the r found by the method of Chapter VII; the original table does not follow sufficiently closely the mathematical form assumed. There is, however, a general difficulty apart from any peculiarity of an individual case, for we can never reach a coefficient of unity because, with a finite number of groups, ϕ^2 can never become infinite which is necessary if r is to be unity. Similarly there is a tendency to mis-state the value of r by the method

* To make this more easy to follow we may mention that the contributions to χ^2 from the first column are 55·1, 3·1, 2·8, 5·8 and 56·0.

† It happens to agree with r from Chapter VII. In the particular case the errors from broad grouping and from deviations from the assumed form happen to balance. The agreement is an illustration of the danger of generalising from isolated cases.

of contingency when r has other values and this depends to some extent on the grouping of the material. Adjustments which are of a fairly simple nature should be made.

11. In §7 when we worked out the connection between r and ϕ^2 we assumed that the frequencies took the form of the normal correlation surface. This means that we assumed that the totals of the columns and rows are "normal curves of error". Let us suppose that we have no finer grouping than that given in the table in §9, then the totals of the columns are 673, 643 1,098 and 456, making a total of 2,870; or reducing them to a total frequency of unity, we have ·2345, ·2240, ·3826 and ·1589. From tables of the "normal curve"* we can work out the ordinates at the end of each group of frequency and form the following table:

Group frequency	(1) for unit frequency n	Total area from beginning (by adding (2))	Ordinate at beginning of group z	Difference of z's negatively	Col (5) Squared	(6)/(2)
(1)	(2)	(3)	(4)	(5)	(6)	(7)
From the columns						
673	·2345	·2345	·00000	− ·30694	·09421	·402
643	·2240	·4585	·30694	− ·08984	·00807	·036
1,098	·3826	·8411	·39678	+ ·15456	·02389	·062
456	·1589	1·0000	·24222	+ ·24222	·05867	·369
...	...	...	·00000	...	...	...
2,870	1·0000	...	...	...	...	·869
					Square root = ·932	
From the rows						
660	·2300	·2300	·00000	− ·30365	·09220	·401
665	·2317	·4617	·30365	− ·09345	·00873	·038
674	·2348	·6965	·39710	+ ·04759	·00226	·010
538	·1875	·8840	·34951	+ ·15421	·02378	·127
333	·1160	1·0000	·19530	+ ·19530	·03814	·329
...	...	...	·00000	...	...	...
2,870	1·0000	...	...	...	...	·905
					Square root = ·951	

* *Tables for Statisticians*, Part I, Table II.

(219)

The final figures are the square roots of

$$\frac{(z_0 - z_1)^2}{n_1} + \frac{(z_1 - z_2)^2}{n_2} + \frac{(z_2 - z_3)^2}{n_3} + \text{etc.}$$

where z_0, z_1, z_2, etc. are the ordinates at the beginning of successive groups and n_1, n_2, n_3, etc. are the proportionate frequencies, i.e. the successive terms in the preceding table col. (2). The corrected value is

$$\frac{\cdot 254}{\cdot 932 \times \cdot 951} = \cdot 286$$

12. The first three columns of the table in the preceding paragraph are easily constructed; the third is wanted because tables of the areas of the "normal curve" give those areas from any point up to the end of the curve, i.e. the integral from x to ∞. The next column gives the ordinate which can be found from the tables in *Tables for Statisticians*, Part i, where the ordinates and areas are in parallel columns, or directly from the tables in Part ii.

We will now turn to the theoretical side and may consider what is the mean of each of the areas n_1, n_2, etc., say, of n_{s+1}.

It will be

$$\int_{x_{s+1}}^{x_s} x e^{-\frac{1}{2}x^2} dx \Big/ \int_{x_{s+1}}^{x_s} e^{-\frac{1}{2}x^2} dx$$

$$= (e^{-\frac{1}{2}x_s^2} - e^{-\frac{1}{2}x_{s+1}^2}) \Big/ \int_{x_{s+1}}^{x_s} e^{-\frac{1}{2}x^2} dx$$

$$= (z_s - z_{s+1})/n_{s+1}$$

Hence this expression gives us the distance of the mean value of the area n_{s+1} from the mean of the whole distribution. But n_{s+1} is the frequency and therefore

$$n_{s+1} \times \left(\frac{z_s - z_{s+1}}{n_{s+1}}\right)^2$$

when summed for all values of s gives the second moment of the distribution and the adjustment, being the square root of a second moment, is a standard deviation.

We had assumed the standard deviations to be unity: we

have now recalculated them on the facts available and adjusted the result. This adjustment in effect removes to a large extent the objections indicated in § 10.

13. It is a little difficult to judge the necessity or success of an adjustment in a case of this kind unless we know the value of the correlation which we ought to reach, and it will probably be more convincing to take one of the coin-tossing tables and, having grouped it, see what values of r are found by the contingency method without adjustment and how near we get to the true value of r with adjustments. For this purpose the table where five coins were common to the pairs of tossings was used and a table was formed as follows:

No. of heads in second tossing	No. of heads in first tossing			Total
	0–4	5–6	7–10	
0–3	2,123 (3,906)	2,541 (1,606)	968 (120)	5,632
4	2,534 (3,360)	3,031 (2,880)	1,155 (480)	6,720
5	3,038 (2,906)	3,640 (4,032)	1,386 (1,126)	8,064
6	2,534 (1,580)	3,031 (3,560)	1,155 (1,580)	6,720
7–10	2,123 (600)	2,541 (2,706)	968 (2,326)	5,632
Total	12,352	14,784	5,632	32,768

The zero-contingency figures are in brackets.

$$\phi^2 = 6971/32768 = \cdot2127$$

$$r \text{ (unadjusted)} = \cdot418$$

Then working the adjustment as before, ·953 was found for the rows and ·892 for the columns, so that the adjusted value of r is $\cdot418/(\cdot953 \times \cdot892) = \cdot492$.

Another trial may be made with the same coin-tossing table throwing it into the form

	0–4	5	6–10	Total
0–4	7,266	2,906	2,180	12,352
5	2,906	2,252	2,906	8,064
6–10	2,180	2,906	7,266	12,352
Total	12,352	8,064	12,352	32,768

Here $\phi^2 = \cdot 171$.

$$r \text{ (unadjusted)} = \cdot 382$$

the factor for rows is ·872 and for columns is the same, so that the adjusted value for r is ·503.

Now both those should be ·5, but clearly ·492 and ·503 are good approximations with broad groupings, and the examples show both the importance of the adjustment and the accuracy attainable.

14. There is, however, one more aspect of this kind of adjustment to which reference may be made. We remarked (§ 8) that the method of contingency implied that we could change the order of the columns and rows; but if we do this, what will happen to the adjustments? The point is of some importance. In broad groups where the division is not quantitative, we may not be sure that if we could express the scale quantitatively it would give a distribution of anything like the assumed normal curve. Let us put this to the test by taking the grouped figures from one of the tables in the preceding paragraph and rearranging them arbitrarily.

Thus we might produce

Second characteristic	FIRST CHARACTERISTIC			Total
	a_0	a_1	a_2	
b_0	4,032	1,126	2,906	8,064
b_1	3,560	1,580	1,580	6,720
b_2	2,880	480	3,360	6,720
b_3	2,706	2,326	600	5,632
b_4	1,606	120	3,906	5,632
Total	14,784	5,632	12,352	32,768

Clearly ϕ^2 and the unadjusted r remain unchanged and so we are only concerned with the totals and the procedure of § 11. Working with these we reach ·944 and ·856 as the factors by which we adjust* and so find a value of ·517 for r. This is better

* The underlying theory of the adjustment is that a normal frequency surface could be cut up to give the table. This could not, I think, be done in the particular rearrangement. But the adjustments work well.

(222)

than the unadjusted value—in fact, quite good. The explanation is that however we divide up the numbers we get adjustments which will not vary to an extreme extent unless the grouping is exceptional.

15. We have already seen that double entry tables will show small values for a measure of correlation even when there is really no correlation and that it is generally more important to decide whether the apparent correlation is significant than to measure exactly the standard error of its coefficient. All we need to do in considering a standard error for ϕ^2 is therefore to compare the actual table with a table formed assuming no correlation and see if the divergence is significant. In practical work it is advisable to make this test before working out the coefficient. Taking, for instance, the table on p. 218, $\chi^2 = 198 \cdot 8$ and we need to find a value for the probability of a divergence as great as or greater than that indicated on the assumption that there is no correlation and that the particular table has arisen merely in sampling.

There are 20 cells in the table but as we fix the totals of each row and each column this would be too large a number to use for n'. The correct number of free cells is $(h-1)(k-1)$ where h is the number of rows and k the number of columns. In the particular case $(h-1)(k-1) = 12$. In *Tables for Statisticians*, Part I, Table XII, n' is used as one more than the number of free cells, i.e. $n' = n + 1$, and we must therefore enter that table with $n' = 13$ and $\chi^2 = 198 \cdot 8$. Knowing the value of r from our previous calculations, it is not surprising to find that the chance of such a divergence from zero correlation is zero to at least six decimal places.

In a fourfold table there is only one free cell.

Another way of setting out the method described in this section is to say that

(i) the mean $\chi^2 = (h-1)(k-1)$ $\qquad\qquad$ ......(9)

(ii) $\sigma_{\chi^2} = \sqrt{\{2(h-1)(k-1)\}}/N$ $\qquad\qquad$ ......(10)

when there is no correlation.

(223)

Formulae (9) and (10) set out the result in a form similar to that already given for η^2 in formulae (4) and (5).

16. In working at contingency we have up to the present assumed that we calculate the squares of the differences between the actual figures in the table and the corresponding figures when there is no correlation, but we may proceed by adding together the differences regardless of sign. We then obtain the mean of these by dividing by the total number of cases. A diagram in *Tables for Statisticians*, Part I, gives values of r. The mathematical work leading to this method is more difficult than that for the mean square contingency given above and in practical work the latter is more dependable.

17. There is yet another method of estimating correlation that may be of help. It is known as correlation of ranks and was suggested by Spearman.* By this method we estimate the correlation between, say, the height and span of a number of schoolchildren without making an exact measurement for any child. We first stand the children in order of height and number them in the rank, the shortest being numbered 1 and the tallest n. Then we rearrange the children in order of span and again number them: the child with the shortest span being numbered 1 and the child with the longest span n. The child numbered 1 in the height rank might be 3 in the span rank and so on.

The next step is to calculate the sum of the squares of the differences between the ranks, say $S(d^2)$, for the two characters (one element in our height and span example would be $(3-1)^2$ or 4) and we can write

$$R = 1 - \frac{6S(d^2)}{n(n^2-1)} \qquad \ldots\ldots(11)$$

$$r = 2\sin\frac{\pi}{6}R \qquad \ldots\ldots(12)$$

where R is the coefficient of correlation between ranks and r

---

* C. Spearman, *Amer. J. Psychol.* xv, 72; K. Pearson, *Drapers' Company Memoir*, No. 4.

the corresponding coefficient between variates—similar to that discussed in previous chapters. The relationship depends on the assumption of the normal correlation surface.

The standard error of r found by this method is approximately 5 per cent. greater than that found by the product-moment method of Chapter X.

CHAPTER XIII

PARTIAL CORRELATION

1. We have up to the present assumed that we can always deal with pairs of related things, but in many investigations, especially perhaps in social statistics, the problems are complicated by a greater number of variables. Suppose, for instance, we were making a study of infant deaths and trying to ascertain the causes chiefly responsible for a high death-rate, we might examine the home environment of children in a particular district to see whether there was any relation between infant deaths and the habits of the mother. But the health of the mother may be important also, and if we find correlation coefficients in respect of (1) infant deaths and habits of mother and (2) infant deaths and health of mother, we have up to the present found no way of eliminating the possible relation between health and habits of the mother. In other words, if the cause of infant mortality is connected with the habits of the mother, is it merely so connected because health and habits are connected?

2. Let us proceed as we did in dealing with correlation where there are only two variables and assume that $x_1 y_1 z_1$, $x_2 y_2 z_2$, etc. be associated deviations, and let

$$z = a + bx + cy \qquad \ldots\ldots(1)$$

As before, we can omit a if we measure every variable from its mean. Then using methods of moments we have

$$(bx_1 + cy_1) + (bx_2 + cy_2) + \ldots = z_1 + z_2 + \ldots$$

$$(bx_1 + cy_1) x_1 + (bx_2 + cy_2) x_2 + \ldots = x_1 z_1 + x_2 z_2 + \ldots$$

or $\qquad\qquad bS(x^2) + cS(xy) = S(xz) \qquad \ldots\ldots(2)$

Similarly $\qquad\quad bS(xy) + cS(y^2) = S(yz) \qquad \ldots\ldots(3)$

(226)

Now, slightly altering the notation used on p. 145, we can write

$$S(xy) = N\sigma_x\sigma_y r_{xy}$$

$$S(x^2) = N\sigma_x^2$$

$$S(y^2) = N\sigma_y^2$$

hence

$$S(xz) = N\sigma_x\sigma_z r_{xz}$$

and

$$S(yz) = N\sigma_y\sigma_z r_{yz}$$

Substituting in (2) and (3), we have

$$b\sigma_x^2 + c\sigma_x\sigma_y r_{xy} = \sigma_x\sigma_z r_{xz}$$

or

$$b\sigma_x + c\sigma_y r_{xy} = \sigma_z r_{xz}$$

and

$$b\sigma_x r_{xy} + c\sigma_y = \sigma_z r_{yz}$$

hence

$$\left. \begin{array}{l} b = \dfrac{\sigma_z}{\sigma_x} \cdot \dfrac{r_{xz} - r_{yz}r_{xy}}{1 - r_{xy}^2} \\[3mm] c = \dfrac{\sigma_z}{\sigma_y} \cdot \dfrac{r_{yz} - r_{xz}r_{xy}}{1 - r_{xy}^2} \end{array} \right\} \quad \ldots\ldots(4)$$

and

Substituting in (1) and remembering that $a = 0$, we have

$$\bar{z} = \frac{\sigma_z}{\sigma_x} \cdot \frac{r_{xz} - r_{yz}r_{xy}}{1 - r_{xy}^2} x + \frac{\sigma_z}{\sigma_y} \cdot \frac{r_{yz} - r_{xz}r_{xy}}{1 - r_{xy}^2} y$$

Now when dealing with the two variables we expressed the result

$$\left. \begin{array}{l} \bar{y} = r\dfrac{\sigma_2}{\sigma_1} x \\[3mm] \bar{x} = r\dfrac{\sigma_1}{\sigma_2} y \end{array} \right\}$$

so that r the measure of the correlation is the geometric mean of the coefficients

$$r\frac{\sigma_2}{\sigma_1} \text{ and } r\frac{\sigma_1}{\sigma_2}$$

(227)

Similarly with three variables we can write down x in terms of z and y or y in terms of x and z, and, again, using the geometric mean of the appropriate pairs of coefficients, we have

$$_yR_{xz} = \frac{r_{xz} - r_{xy}r_{yz}}{\sqrt{\{(1 - r_{xy}^2)(1 - r_{yz}^2)\}}}$$

as the net (or partial) coefficient between x and z associated with a single type of y. The square root in the denominator is to be taken as positive.

3. Now coefficients of correlation must not exceed unity, therefore

$$(r_{xz} - r_{xy}r_{yz})^2 \not> (1 - r_{xy}^2)(1 - r_{yz}^2)$$

or, r_{xz} must lie between the limits

$$r_{xy}r_{yz} \pm \sqrt{\{(1 - r_{xy}^2)(1 - r_{yz}^2)\}}$$

From this we can write down some of the limits that may arise when we are dealing with three variables.

If		Then
$r_{xy} =$	$r_{yz} =$	$r_{xz} =$
0	0	any value
1	1	1
-1	-1	1
1	-1	-1
0	± 1	0
0	$\pm r$	between $\pm \sqrt{(1 - r^2)}$
r	r	between 1 and
$-r$	$-r$	$2r^2 - 1$
r	$-r$	between $1 - 2r^2$ and -1

4. We may now consider the following numerical example:*

* "Relative value of factors influencing infant welfare", *Annals of Eugenics*, I, 178–9. The statistics quoted are from Bradford, 1911. The student will find many similar sets of tables in this paper.

Habits of Mother (x)	Health of Mother (y)		Total
	Good	Not good	
Good	956	197	1,153
Indifferent	257	286	543
Total	1,213	483	1,696

$$r_{xy} = \cdot567 \pm \cdot033$$

Child dead or not (z)	Habits of Mother (x)		Total
	Good	Indifferent	
Living	997	420	1,417
Dead	156	123	279
Total	1,153	543	1,696

$$r_{xz} = \cdot213 \pm \cdot046$$

Child dead or not (z)	Health of Mother (y)		Total
	Good	Not good	
Living	1,065	352	1,417
Dead	148	131	279
Total	1,213	483	1,696

$$r_{yz} = \cdot329 \pm \cdot045$$

Let us now work out our partial coefficient between "Habits of Mother" and "Infantile Deaths" for constant "Health of Mother", and we have

$$_{y}R_{xz} = \frac{\cdot213 - (\cdot329)(\cdot567)}{\sqrt{(\cdot891)}\sqrt{(\cdot678)}} = \cdot034$$

In other words the value, though it looked like ·213 at first, now proves to be only ·034, and as the standard error is about ·04 we could not say that the result is significant.

If we worked out the partial coefficient between "Health of Mother" and "Infant Deaths", keeping "Habits" constant, we reach a value of ·26, which is significant though smaller than the crude figure of · 329.

5. It is possible to extend the theory to a larger number of variables, but it seems unnecessary to do so here. The example will give an indication of the use to which such work may be put, and supplies a warning against accepting the numerical value of a coefficient until other causes that may affect the result have been considered.

APPENDIX I

CORRECTIONS FOR MOMENTS

1. The following method has been suggested by E. Pairman and K. Pearson (*Biometrika*, XII, 231 et seq.) when the curve rises abruptly at one or both ends.

Let n_1, n_2, etc. be the proportionate frequencies in the 1st, 2nd, etc. groups, then put

$$a_1 = -\tfrac{1}{60}\{137n_1 - 163n_2 + 137n_3 - 63n_4 + 12n_5\}$$

$$a_2 = \tfrac{1}{12}\{45n_1 - 109n_2 + 105n_3 - 51n_4 + 10n_5\}$$

$$a_3 = -\tfrac{1}{4}\{17n_1 - 54n_2 + 64n_3 - 34n_4 + 7n_5\}$$

$$a_4 = \{3n_1 - 11n_2 + 15n_3 - 9n_4 + 2n_5\}$$

$$a_5 = -\{n_1 - 4n_2 + 6n_3 - 4n_4 + n_5\}$$

Similarly, values of b_1, b_2 etc. can be obtained from the other end of the distribution.

Then the values of the moments are as follows, where A is the distance of the start and B is the distance of the end of the distribution from the origin about which moments are calculated:

$$\mu_1' = \nu_1' + \{\tfrac{1}{12}(a_1 - \tfrac{1}{60}a_3 + \tfrac{1}{2520}a_5) + \tfrac{1}{12}(b_1 - \tfrac{1}{60}b_3 + \tfrac{1}{2520}b_5)\}$$

$$\mu_2' = \nu_2' - \tfrac{1}{12} + \{-\tfrac{1}{120}(a_2 - \tfrac{5}{126}a_4) + \tfrac{1}{6}A(a_1 - \tfrac{1}{60}a_3 + \tfrac{1}{2520}a_5)\}$$

$$\mu_3' = \nu_3' - \tfrac{1}{4}\nu_1' + \{-\tfrac{1}{40}(a_1 - \tfrac{5}{63}a_3 + \tfrac{1}{240}a_5)$$
$$- \tfrac{1}{40}A(a_2 - \tfrac{5}{126}a_4) + \tfrac{1}{4}A^2(a_1 - \tfrac{1}{60}a_3 + \tfrac{1}{2520}a_5)\}$$

$$\mu_4' = \nu_4' - \tfrac{1}{2}\nu_2' + \tfrac{7}{240} + \{\tfrac{1}{126}(a_2 - \tfrac{7}{80}a_4) - \tfrac{1}{10}A(a_1 - \tfrac{5}{63}a_3 + \tfrac{1}{240}a_5)$$
$$- \tfrac{1}{20}A^2(a_2 - \tfrac{5}{126}a_4) + \tfrac{1}{3}A^3(a_1 - \tfrac{1}{60}a_3 + \tfrac{1}{2520}a^5)\}$$

and similar expressions in B and b's.

If the moments be taken about the start of the first group so that the first group is multiplied by powers of $\frac{1}{2}$, the second by powers of $\frac{3}{2}$ and so on, this expression is simplified so far as the a terms are concerned because the terms involving A vanish.

2. The method of reaching these adjustments starts with the Euler-Maclaurin expansion and assumes that the curve takes the form

$$1 + \frac{a_1}{1!}(x - A) + \frac{a_2}{2!}(x - A)^2, \text{ etc.}$$

at the beginning and a similar form at the end. This leads to the values of the a's.

The differential coefficients at each end required in the Euler-Maclaurin expansion are then evolved and the result given is reached.

The frequency at the start is approximately

$$\frac{N}{60}\{137n_1 - 163n_2 + 137n_3 - 63n_4 + 12n_5\}$$

By means of this expression we can discover how nearly the frequency curve comes to zero at the ends of the range.

3. A few numerical examples may be given. The rule that the area, in the case of high contact, can be found by adding ordinates when tested by adding 12 ordinates of the normal curve calculated to 5 decimal places, gave 1·24998 instead of 1·25000. Nine ordinates of a Type III curve with high contact gave 24473 instead of 24475.

An example of the method of § 1 above is taken from the paper there cited. Moments for $\sqrt{x} \times 100{,}000$ from $x = 0$ to $x = 10$ were calculated, the exact result being known. The proportional frequencies, which may be taken as the data, were:

$n_1 = $ ·031623	$n_6 = $ ·111205
$n_2 = $ ·057820	$n_7 = $ ·120904
$n_3 = $ ·074874	$n_8 = $ ·129880
$n_4 = $ ·088665	$n_9 = $ ·138273
$n_5 = $ ·100571	$n_{10} = $ ·146185

$$1{\cdot}000000$$

From these figures

$$a_1 = -\cdot 0131,0643 \qquad b_1 = \cdot 1499,9857$$
$$a_2 = -\cdot 0444,8167 \qquad b_2 = -\cdot 0074,9283$$
$$a_3 = \cdot 0258,4150 \qquad b_3 = -\cdot 0003,9450$$
$$a_4 = -\cdot 0148,8400 \qquad b_4 = -\cdot 0000,2600$$
$$a_5 = \cdot 0045,0200 \qquad b_5 = -\cdot 0000,3800$$

$$a_1 - \tfrac{1}{60}a_3 + \tfrac{1}{2520}a_5 = -\cdot 0135,3533$$
$$a_2 - \tfrac{5}{126}a_4 = -\cdot 0438,9104 \text{ and so on.}$$

Putting $A = 0$ and $B = 10$ and calculating moments about the start, we require for the a adjustments

$$\tfrac{1}{12}(a_1 - \tfrac{1}{60}a_3 + \tfrac{1}{2520}a_5) = -\cdot 0011,2794$$

and the other adjustments in order are

$$-\cdot 0003,6576, \quad -\cdot 0003,7846 \text{ and } -\cdot 0003,4269$$

For the b terms we have

$$\tfrac{1}{12}(b_1 - \tfrac{1}{60}b_3 + \tfrac{1}{2520}b_5) = \cdot 0121,0043$$
$$\tfrac{1}{6}B(b_1 - \tfrac{1}{60}b_3 + \tfrac{1}{2520}b_5) = \cdot 2500,0860$$

and the other terms in order give

$$3\cdot 7501,2825, \qquad 50\cdot 0017,1000$$
$$-\cdot 0000,6244, \qquad -\cdot 0001,8731$$
$$-\cdot 0374,6365, \qquad \cdot 0037,5074$$
$$\cdot 1500,2972, \qquad -\cdot 0000,5945$$

Finally for the adjusted moments we reach

	Raw moments	With Sheppard's adjustment	With full adjustment	True value	
ν'_1	5·9880	5·9880	5·9994	6·0000	μ'_1
ν'_2	42·6900	42·6067	42·8570	42·8571	μ'_2
ν'_3	331·0854	329·5884	333·3349	333·3333	μ'_3
ν'_4	2698·7735	2677·4576	2727·2757	2727·2727	μ'_4

4. The method described above gives good results but is laborious. The approximations are less satisfactory in those cases where the first group does not relate to a complete unit

base and the curve rises abruptly. The same authors gave a method for J-shaped distributions, but I should not use it as a simple approximation can be found by examining the exponential (Type X).

When statistics expressible by the exponential $y = y_0 e^{-x/\sigma}$ are stated in groups for each equal subrange h of x, the successive groups are $\int_0^h y_0 e^{-x/\sigma} dx; \int_h^{2h} y_0 e^{-x/\sigma} dx$; etc.; or

$$y_0 \sigma (1 - e^{-h/\sigma}); \quad y_0 \sigma (1 - e^{-h/\sigma}) e^{-h/\sigma}; \quad y_0 \sigma (1 - e^{-h/\sigma}) e^{-2h/\sigma}; \text{ etc.}$$

These terms may also be regarded as a geometrical progression,* the first term being $y_0 \sigma (1 - e^{-h/\sigma})$ and the common ratio $e^{-h/\sigma}$. It follows that if we treat the areas as a geometrical progression extending to infinity, calculate the moments on this assumption and read the result as graduated terms of a geometrical progression, we shall reach correctly graduated areas, and we can subsequently write down the equation to the curve with little trouble.

Other points are however involved. Let us write the geometrical progression as ka^x and put $A = (1 - a)^{-1}$, then the moments about its mean are

2nd moment $A^2 - A$

3rd   ,, $2A^3 - 3A^2 + A$

4th   ,, $9A^4 - 18A^3 + 10A^2 - A$

and if we work out β_1 and β_2, we get $4 + h^2/\mu_2$ and $9 + h^2/\mu_2$ respectively.

Using the exponential, the moments, etc. about the mean are: $\mu_2 = \sigma^2$, $\mu_3 = 2\sigma^3$, $\mu_4 = 9\sigma^4$, $\beta_1 = 4$, $\beta_2 = 9$.

Hence when we calculate moments, assuming that the statistics form a geometrical progression, whereas they are really areas from a curve, and seek to choose the type of curve from Pearson's criteria in his system, we shall reach a persistent error. For this purpose the β_1 and β_2 found from the statistics should be reduced by h^2/μ_2.

* "Geometrical progression" is used throughout to describe a discrete series and exponential curve to describe a continuous one.

This rule can be used as an approximation in all J-shaped curves and will be found to give satisfactory results.

So far we have assumed that we know the start of the curve and that all the bases of the areas are of equal size. If this does not apply we can, in the case of an exponential curve, fit the curve, excluding the first (incomplete) term, and regard that term as related to an appropriate base extrapolated from the graduation of the remainder. This is an arbitrary arrangement but has practical advantages.

In other J-shaped curves in similar circumstances the first step would be to assume an exponential, to find therefrom approximately the base of the first incomplete group, and then assume that the area is concentrated at the middle point. This will generally give good results: the assumption of the exponential overstates the base and the assumption of half-way assumes a less rapidly falling curve than the J-shaped forms of Types I and III. There is therefore a balance of error.

Turning to the statistical side, the example on p. 108 gives $\mu_2 = 2 \cdot 045$, $\beta_1 = 4 \cdot 629$, $\beta_2 = 9 \cdot 502$. These figures come from the unadjusted moments, and deducting $\cdot 49$ from the above values for β_1 and β_2 we reach $4 \cdot 14$ and $9 \cdot 01$. The theoretical values when an exponential curve is to be used are 4 and 9.

If we apply the rule as an approximation in other J-shaped cases we find that in the example on p. 112, where a twisted J-shaped curve is given, $\mu_2 = 4 \cdot 266$, $\beta_1 = \cdot 761$, $\beta_2 = 2 \cdot 646$, and the adjustment leads to $\beta_1 = \cdot 527$ and $\beta_2 = 2 \cdot 412$. Hence $5\beta_2 - 6\beta_1 - 9$ becomes $- \cdot 098$ instead of $- \cdot 368$. The theoretical criterion would lead us to expect $5\beta_2 - 6\beta_1 - 9 = 0$.

These examples are not, of course, complete evidence, but they show that the suggestion may lead to accurate results, and it has the merit of simplicity. The rule with regard to the adjustment of the β's by h^2/μ_2 may be combined with the approximations given on p. 109, where it is mentioned that the mean is overstated, when μ_3 is positive, by about $\dfrac{h^2}{12\sigma}$, and the second moment about the *true* mean (i.e. the mean as corrected by $h^2/(12\sigma)$) is understated by about $\dfrac{h^2}{12}$. We do not know σ exactly but can use the square root of the second moment as found from the calculations. If h be taken as a unit and the moments found in terms of h, i.e. in working units, the corrections are $1/(12\sqrt{\mu_2})$ and $\tfrac{1}{12}$.

5. An alternative to the method of §1 is to find mid-ordinates corresponding to the areas of the groups and treat these mid-ordinates in the manner explained in Chapter III, §18.

The mid-ordinates m_1, m_2, etc. are found by the following equations:

$$m_3 = \tfrac{1}{1920}\{2134n_3 - 116(n_2 + n_4) + 9(n_1 + n_5)\}$$
$$m_2 = \tfrac{1}{1920}\{-71n_1 + 2044n_2 - 26n_3 - 36n_4 + 9n_5\}$$
$$m_1 = \tfrac{1}{1920}\{1689n_1 + 684n_2 - 746n_3 + 364n_4 - 71n_5\}$$

The total frequency is not exactly reproduced but the moments obtained are good approximations.

6. It has been pointed out that one of the difficulties in calculating moments when the curve rises abruptly at one or both ends arises because the true start or end of the curve is unknown. In other words, the base of the first area or last area (or both) is smaller than that of the other areas. In practice good results can often be obtained with unadjusted moments but the first attempt may require modification by varying the range of the curve (see p. 124). When this is done, the moment contribution for the first area, or the last, or both, must be recalculated by assuming that the area is concentrated at the middle of the smaller base.

E. S. Martin has approached the problem more systematically in a paper in *Biometrika*, XXIV, 12, and has given tables from which the start of the curve may be estimated.

APPENDIX II

B AND $\varGamma$ FUNCTIONS

$$B(m, n) = \int_0^1 x^{m-1} (1 - x)^{n-1} dx$$

$$\varGamma(p) = \int_0^\infty e^{-x} x^{p-1} dx$$

I. $\quad \int x^{p-1} e^{-x} dx = -e^{-x} x^{p-1} + (p - 1) \int x^{p-2} e^{-x} dx$

by integration by parts.

When $p - 1$ is positive, $e^{-x} x^{p-1}$ vanishes when $x = 0$, and when $x = \infty$ it can be written x^{p-1}/e^x, and the rule for evaluation of undetermined forms can be applied giving zero as the result.

$$\int_0^\infty x^{p-1} e^{-x} dx = (p - 1) \int_0^\infty x^{p-2} e^{-x} dx$$

or $\qquad \varGamma(p) = (p - 1) \varGamma(p - 1)$

If p be an integer, $\varGamma(p) = (p - 1)!$

II. To prove $B(m, n) = \dfrac{\varGamma(m) \varGamma(n)}{\varGamma(m + n)}$.

Putting zx for x in the equation for $\varGamma(m)$, we have

$$\varGamma(m) = \int_0^\infty e^{-zx} z^m x^{m-1} dx$$

and $\qquad \varGamma(m) e^{-z} z^{n-1} = \int_0^\infty e^{-z(1+x)} z^{m+n-1} x^{m-1} dx$

$$\therefore \quad \varGamma(m) \int_0^\infty e^{-z} z^{n-1} dz = \int_0^\infty \left[\int_0^\infty e^{-z(1+x)} z^{m+n-1} dz \right] x^{m-1} dx$$

(237)

But if $z(1+x) = y$, we get

$$\int_0^\infty e^{-z(1+x)} z^{m+n-1} dz = \frac{1}{(1+x)^{m+n}} \int_0^\infty e^{-y} y^{m+n-1} dy$$

$$= \frac{\Gamma(m+n)}{(1+x)^{m+n}}$$

$\therefore \qquad \Gamma(m)\,\Gamma(n) = \Gamma(m+n) \int_0^\infty \frac{x^{m-1}}{(1+x)^{m+n}}\, dx$

But putting $1+x = \dfrac{1}{1-z}$ in this integral, we obtain

$$\int_0^1 \frac{z^{m-1}}{(1-z)^{m-1}} (1-z)^{m+n} \frac{1}{(1-z)^2}\, dz$$

which reduces at once to $B(m, n)$, and therefore

$$B(m, n) = \frac{\Gamma(m)\,\Gamma(n)}{\Gamma(m+n)}$$

III. To prove $\Gamma(\tfrac{1}{2}) = \sqrt{\pi}$.

We have already shown (see p. 84) that $2 \displaystyle\int_0^\infty e^{-x^2} dx = \sqrt{\pi}$, and by putting $x^2 = z$, we have

$$2 \int_0^\infty e^{-x^2} dx = \int_0^\infty e^{-z} z^{-\frac{1}{2}} dz = \Gamma(\tfrac{1}{2}) = \sqrt{\pi}$$

For statistical work a table of $\Gamma(x)$ or $\log \Gamma(x)$ is required (see pp. 266–7 or *Tables for Statisticians*, or, better still, J. Brownlee, "Log $\Gamma(x)$ from $x = 1$ to 50·9 by intervals of 0·01", *Tracts for Computers*, No. IX, Camb. Univ. Press). Tracts Nos. IV and VIII give other values of Γ functions.

When x is large, we can be approximate to $\log \Gamma(x)$. The best known approximation is

$$\Gamma(x+1) = \sqrt{(2\pi x)}\, x^x\, e^{-x}\, e^{1/12x}*$$

or

$$\log_{10} \Gamma(x+1) = \log_{10} \sqrt{(2\pi)} + (x + \tfrac{1}{2}) \log_{10} x - \left(x - \frac{1}{12x}\right) \log_{10} e$$

---

* A proof of this well-known approximation will be found in *Chrystal's Algebra*, II, 368, etc., or in *Boole's Finite Differences*, Chapter VI.

and it can be used when x is not less than 8. The following calculations show how the table of log $\Gamma(x)$ is used, and how the approximation approaches the true value.

x	Log $\Gamma(x)$		
	True	Approximate	Δ
1·372	1·948975	...	...
2·372	·086329	·086743	·000414
3·372	·461444	·461532	·000088
4·372	·989332	·989369	·000037
5·372	1·630012	1·630025	·000013
6·372	2·360148	2·360156	·000008
7·372	3·164424	3·164430	·000006
8·372	4·032009	4·032010	·000001
9·372	4·954838	4·954837	− ·000001
10·372	5·926670	5·926669	− ·000001

The functions already described may be described as "complete Γ-functions" and "complete B-functions" but for certain purposes "incomplete" functions are required. Thus, if we have to express the areas of a Type III curve we require $\int_a^b e^{-x}x^p \, dx$ where the base is $b - a$, and these integrals can be obtained by subtracting $\int_0^a e^{-x}x^p \, dx$ from $\int_0^b e^{-x}x^p \, dx$. The incomplete Γ-function, $\Gamma_x(p+1)$, is defined as $\int_0^x e^{-x}x^p \, dx$. Bearing in mind that y_0 involves the corresponding complete Γ-function in the denominator and the values assumed by Γ-functions, it has been found preferable to tabulate a function called $I(x, p)$ which is $\Gamma_x(p+1)/\Gamma(p+1)$. This function always lies between 0 and 1, and expresses the chance of a variation within a given limit if the frequency is expressible by a Type III curve. There is however a difficulty in tabulation which has been met by using, instead of x, the argument $u = x/\sqrt{(p+1)}$. Tables of $I(u, p)$ have been published.*

* *Tables of the Incomplete Γ-Function.* Printed by Cambridge University Press and published by *Biometrika*.

(239)

The incomplete B-function is defined as

$$B_x(p, q) = \int_0^x x^{p-1} (1-x)^{q-1} dx$$

and it is tabulated and published* in the form

$$I_x(p, q) = B_x(p, q)/B(p, q)$$

Though the tables of the incomplete Γ and B functions are valuable for certain purposes, the student will find that for ordinary curve-fitting the methods described in Chapter V will suffice.

* *Tables of the Incomplete Beta-Function.* Published by *Biometrika*.

APPENDIX III

THE EQUATION TO THE NORMAL SURFACE

Let $\eta_1, \eta_2, \eta_3, \ldots \eta_n$ be deviations from their respective means of a complex of measurable characteristics. The sizes of the functions measured, or organs, are determined by a large number of *independent* contributory causes. Let there be m of these causes, and let their deviations from their means be $\epsilon_1, \epsilon_2, \epsilon_3, \ldots \epsilon_m$, then $\eta_1, \eta_2, \eta_3, \ldots \eta_n$ will be functions of $\epsilon_1, \epsilon_2, \epsilon_3, \ldots \epsilon_m$. Further, if $m > n$ certain of the ϵ's will appear only in certain of the η's, and the ϵ's will not be fully determined for a given η complex. We also assume that the variations in intensity of the contributory causes are small as compared with their absolute intensity, and that these variations follow the normal law of distribution; that is, we assume that the deviations from the mean value can be graduated by the normal curve of error. The mean complex being reached with the mean intensities of contributory causes, we have, by the principle of the super position of small quantities,

$$
\left.
\begin{aligned}
\eta_1 &= a_{11}\epsilon_1 + a_{12}\epsilon_2 + a_{13}\epsilon_3 + \ldots + a_{1m}\epsilon_m \\
\eta_2 &= a_{21}\epsilon_1 + a_{22}\epsilon_2 + a_{23}\epsilon_3 + \ldots + a_{2m}\epsilon_m \\
&\cdots\cdots\cdots\cdots\cdots\cdots\cdots\cdots\cdots\cdots\cdots\cdots\cdots\cdots\cdots\cdots \\
\eta_n &= a_{n1}\epsilon_1 + a_{n2}\epsilon_2 + a_{n3}\epsilon_3 + \ldots + a_{nm}\epsilon_m
\end{aligned}
\right\} \quad \ldots\ldots(\text{i})
$$

The a's are coefficients whose values have to be determined, and any of the system of a's may be zero, for a particular contributory cause may have no effect on a particular result. Further, the chance that we have a conjunction of contributory causes lying between ϵ_1 and $\epsilon_1 + \delta\epsilon_1$, ϵ_2 and $\epsilon_2 + \delta\epsilon_2, \ldots$ and between ϵ_m and $\epsilon_m + \delta\epsilon_m$ will be given by

$$
P' = K e^{-\left(\frac{\epsilon_1^2}{2\kappa_1^2} + \frac{\epsilon_2^2}{2\kappa_2^2} + \ldots + \frac{\epsilon^2_m}{2\kappa^2_m}\right)} \times \delta\epsilon_1 \delta\epsilon_2 \ldots \delta\epsilon_m \quad \ldots\ldots(\text{ii})
$$

where the standard deviations of the distributions are $\kappa_1, \kappa_2, \ldots \kappa_m$ and K is constant.*

Now, by (i) let n of the variables ϵ, say the first n, be replaced by the variables η, then the probability that we have a complex with organs lying between η_1 and $\eta_1 + \delta\eta_1$, η_2 and $\eta_2 + \delta\eta_2, \ldots \eta_n$ and $\eta_n + \delta\eta_n$, together with a series of contributory causes lying between ϵ_{n+1} and $\epsilon_{n+1} + \delta\epsilon_{n+1}$, ϵ_{n+2} and $\epsilon_{n+2} + \delta\epsilon_{n+2}, \ldots \epsilon_m$ and $\epsilon_m + \delta\epsilon_m$, will be

$$P' = C'e^{-\frac{1}{2}\phi^2}\delta\eta_1\delta\eta_2 \ldots \delta\eta_n\delta\epsilon_{n+1}\delta\epsilon_{n+2} \ldots \delta\epsilon_m$$

where C' is a constant, a function of K and the a's, and ϕ^2 consists of the following parts:

(i) A quadratic function of the η's from η_1 to η_n.

(ii) A quadratic function of the ϵ's from ϵ_{n+1} to ϵ_m.

(iii) A series of functions of the type

$$\epsilon_{n+1}(b_{1, n+1}\eta_1 + b_{2, n+1}\eta_2 + \ldots + b_{n, n+1}\eta_n)$$
$$\epsilon_{n+1}(b_{1, n+2}\eta_1 + b_{2, n+2}\eta_2 + \ldots + b_{n, n+2}\eta_n)$$
$$\ldots\ldots\ldots\ldots\ldots\ldots\ldots\ldots\ldots\ldots\ldots\ldots\ldots\ldots\ldots\ldots\ldots$$
$$\epsilon_m(b_{1, m}\eta_1 + b_{2, m}\eta_2 + \ldots + b_{n, m}\eta_n)$$

where some of the b's may be zero.

Now, if P' be integrated for all values from $-\infty$ to $+\infty$ of all the contributory causes $\epsilon_{n+1}, \epsilon_{n+2}, \ldots \epsilon_m$, we shall have the whole chance of a complex with organs falling between η_1 and $\eta_1 + \delta\eta_1$, η_2 and $\eta_2 + \delta\eta_2, \ldots \eta_n$ and $\eta_n + \delta\eta_n$. But every time we integrate with regard to an ϵ, say ϵ_{n+1}, we alter the constants of each contributory part of ϕ^2, but do not alter the triple constitution of ϕ^2 except to cause one ϵ to disappear from its (ii) and (iii) constituents. At the same time we alter C' without introducing into it any terms in η. Thus, finally, after $m - n$ integrations, ϕ^2 is reduced to its first constituent, or we conclude that the

---

* Considering any particular case of the normal curves, the chance of getting a result between ϵ_1 and $\epsilon_1 + \delta\epsilon_1$ when the distribution is "normal" is $y_0 e^{-\epsilon_1^2/2\kappa_1^2}\delta\epsilon_1$ where κ_1 is the standard deviation; similarly with each of the other causes. As the causes are independent, the product of the various chances gives the required chance.

chance of a complex of organs between η_1 and $\eta_1 + \delta\eta_1$, η_2 and $\eta_2 + \delta\eta_2, \dots \eta_n$ and $\eta_n + \delta\eta_n$ occurring is given by

$$P = Ce^{-\frac{1}{2}\chi^2}\delta\eta_1\,\delta\eta_2\,\delta\eta_3 \dots \delta\eta_n \qquad \dots\dots(\text{iii})$$

where χ^2 is a quadratic function of the η's. This is the law of frequency for the complex.

Consider the expression (iii) but replace χ^2 by a quadratic function, then

$$P = Ce^{-\frac{1}{2}\{c_{11}\eta_1{}^2 + c_{22}\eta_2{}^2 + \dots + 2c_{12}\eta_1\eta_2 + 2c_{13}\eta_1\eta_3 + \dots\}}$$
$$= Ce^{-\frac{1}{2}\{S_1(c_{pp}\eta_p{}^2) + 2S_2(c_{pq}\eta_p\eta_q)\}}$$

Here C, c_{pp}, c_{pq} are constants, and S_1 denotes a summation for every value of p, and S_2 for every pair of values of p and q in the series.

Taking the simplest case of two variables, this becomes, with slight modifications of notation,

$$P = Ce^{-\frac{1}{2}(c_1\eta_1{}^2 + c_2\eta_2{}^2 + 2c_{12}\eta_1\eta_2)}$$

Integrate P for all values of η_1 from $-\infty$ to $+\infty$, and we must have the normal curve of η_2 variation.* Therefore

$$\frac{1}{2\sigma_1^2} = c_2\left(1 - \frac{c_{12}^2}{c_1 c_2}\right)$$

Similarly, integrating for all values of η_2,

$$\frac{1}{2\sigma_2^2} = c_1\left(1 - \frac{c_{12}^2}{c_1 c_2}\right)$$

Integrating for all values of η_1 and η_2 to obtain the total frequency, we have

$$N = \frac{2C\pi}{\sqrt{(c_1 c_2 - c_{12}^2)}}$$

Now put $r = -\dfrac{c_{12}}{\sqrt{(c_1 c_2)}}$ and write x and y for η_1 and η_2, and we have

$$Z = \frac{N}{2\pi\sigma_1\sigma_2\sqrt{(1-r^2)}} e^{-\frac{1}{2}\left\{\frac{x^2}{\sigma_1{}^2(1-r^2)} - \frac{2xyr}{\sigma_1\sigma_2(1-r^2)} + \frac{y^2}{\sigma_2{}^2(1-r^2)}\right\}}$$

$$\dots\dots(\text{iv})$$

* The result $\dfrac{1}{2\sigma_1{}^2} = c_2\left(1 - \dfrac{c_{12}{}^2}{c_1 c_2}\right)$ can be reached at once by re-arranging the index in the expression for P as $\left\{\text{a perfect square} - \frac{1}{2}c_2\left(1 - \dfrac{c_{12}{}^2}{c_1 c_2}\right)\eta_2{}^2\right\}$.

APPENDIX IV

THE INTEGRATION OF SOME EXPRESSIONS CONNECTED WITH THE NORMAL CURVE OF ERROR

1. On p. 84 we showed that

$$\int_{-\infty}^{\infty} e^{-x^2} dx = \sqrt{\pi} \qquad \qquad \dots\dots(i)$$

$$\therefore \qquad \int_{-\infty}^{\infty} e^{-x^2/h^2} dx = h\sqrt{\pi} \qquad \qquad \dots\dots(ii)$$

Since the curve is symmetrical, we have

$$\int_{-\infty}^{+\infty} x^{2n+1} e^{-x^2} dx = 0 \qquad \qquad \dots\dots(iii)$$

If we integrate $\int x^{2n} e^{-x^2} dx$ by parts, we have

$$\int x^{2n} e^{-x^2} dx = \frac{x^{2n+1}}{2n+1} e^{-x^2} + \int 2x \frac{x^{2n+1}}{2n+1} e^{-x^2} dx$$

and inserting the limits $-\infty$ and $+\infty$

$$\int_{-\infty}^{+\infty} x^{2n} e^{-x^2} dx = \frac{2}{2n+1} \int_{-\infty}^{+\infty} x^{2n+2} e^{-x^2} dx \qquad \dots\dots(iv)$$

This last formula shows the connection between the successive even moments.

2. Let

$$Z = Z_0 e^{-\frac{1}{2(1-r^2)}\left\{\frac{x^2}{\sigma_1^2} - \frac{2xyr}{\sigma_1\sigma_2} + \frac{y^2}{\sigma_2^2}\right\}}$$

Then Z can be put in the form

$$Z_0 e^{-\frac{1}{2(1-r^2)}\left\{\frac{x}{\sigma_1} - \frac{yr}{\sigma^2}\right\}^2} e^{-\frac{1}{2(1-r^2)}\cdot\frac{y^2(1-r^2)}{\sigma_2^2}} = Z_0 e^{-\frac{1}{2(1-r^2)\sigma_1^2}\left\{x - \frac{yr\sigma_1}{\sigma_2}\right\}^2} e^{-\frac{y^2}{2\sigma_2^2}}$$

Then $\qquad \int_{-\infty}^{+\infty} Z\, dx = Z_0 \sqrt{\{2\pi(1-r^2)\}} \,\sigma_1 e^{-y^2/2\sigma_2^2}$ by (ii)

and

$$N = \int_{-\infty}^{+\infty}\int_{-\infty}^{+\infty} Z\,dx\,dy = \int_{-\infty}^{+\infty} Z_0\sqrt{\{2\pi(1-r^2)\}}\,\sigma_1 e^{-y^2/2\sigma_2^2}\,dy$$

$$= 2\pi\sigma_1\sigma_2\sqrt{(1-r^2)}\,Z_0 \qquad \ldots\ldots(\mathrm{v})$$

3. Using the same method as that just given, it can be shown that if $ac > b^2$

$$\int_{-\infty}^{+\infty}\int_{-\infty}^{+\infty} e^{-\frac{1}{2}(ax^2-2bxy+cy^2)}\,dx\,dy = \frac{2\pi}{\sqrt{(ac-b^2)}} \qquad \ldots\ldots(\mathrm{vi})$$

for the index can be written

$$-\frac{a}{2}\left\{x-\frac{b}{a}y\right\}^2 + \frac{y^2}{2a}(ac-b^2)$$

and if we then integrate with respect to x, we have

$$\sqrt{\frac{2\pi}{a}}\,e^{-\frac{y^2}{2a}(ac-b^2)}$$

and if $ac - b^2$ is positive, we can integrate this last expression with respect to y and have

$$\sqrt{\frac{2\pi}{a}}\sqrt{\frac{2\pi a}{ac-b^2}} = \frac{2\pi}{\sqrt{(ac-b^2)}}$$

4. We shall now find $\int_{-\infty}^{+\infty}\int_{-\infty}^{+\infty} Zxy\,dx\,dy.$

Proceeding as in (v), we have

$$\int_{-\infty}^{+\infty} Zxy\,dx = \int_{-\infty}^{+\infty} xy Z_0 e^{-y^2/2\sigma_2^2}\,e^{-\frac{1}{2(1-r^2)\sigma_1^2}\left\{x-\frac{yr\sigma_1}{\sigma_2}\right\}^2}\,dx$$

$$= Z_0 y e^{-y^2/2\sigma_2^2}\int_{-\infty}^{+\infty}\left(X+\frac{yr\sigma_1}{\sigma_2}\right)e^{-\frac{X^2}{2(1-r^2)\sigma_1^2}}\,dX$$

$$\text{where } X = x - \frac{yr\sigma_1}{\sigma_2}$$

$$= Z_0 y e^{-y^2/2\sigma_2^2}\int_{-\infty}^{+\infty}\frac{yr\sigma_1}{\sigma_2}\,e^{-\frac{X^2}{2(1-r^2)\sigma_1^2}}\,dx$$

$$\text{because by (iii) } \int_{-\infty}^{+\infty} e^{-X^2/h^2} X\,dX \text{ is zero}$$

$$= Z_0\frac{\sigma_1^2}{\sigma_2}r\sqrt{\{2\pi(1-r^2)\}}\,y^2 e^{-y^2/2\sigma_2^2}$$

(245)

But, by putting $n = 0$ in (iv) and using (ii), we have

$$\int_{-\infty}^{+\infty} y^2 e^{-y^2/2\sigma_2^2}\, dy = \sigma_2^3 \sqrt{(2\pi)}$$

$$\therefore \qquad \int_{-\infty}^{+\infty}\int_{-\infty}^{+\infty} Zxy\, dx\, dy = Z_0 \sigma_1^2 \sigma_2^2\, 2\pi r\, \sqrt{(1-r^2)}$$

$$= N\sigma_1\sigma_2 r \qquad \ldots\ldots\text{(vii)}$$

because by (v)

$$Z_0 = \frac{N}{2\pi\sigma_1\sigma_2\sqrt{(1-r^2)}}$$

5. We may now deal with the problem referred to in Chapter IX. "To find a value for r from the equation

$$\frac{N}{2\pi\sqrt{(1-r^2)}}\int_h^\infty\int_k^\infty e^{-\frac{1}{2}\frac{1}{1-r^2}(x^2+y^2-2rxy)}\, dx\, dy = d$$

where d, N, h, and k are known."

Consider the expression

$$\frac{1}{\sqrt{(1-r^2)}}\, e^{-(x^2+y^2-2rxy)/2(1-r^2)} = U,\ \text{say} \qquad \ldots\ldots(\alpha)$$

and expand it in terms of r by Maclaurin's theorem, then

$$U = e^{-\frac{1}{2}(x^2+y^2)}\left(u_0 + \frac{u_1 r}{1!} + \frac{u_2 r^2}{2!} + \ldots + \frac{u_n r^n}{n!} + \ldots\right) \ldots\ldots(\beta)$$

where $u_n = e^{\frac{1}{2}(x^2+y^2)}\left(\dfrac{d^n U}{dr^n}\right)_{r=0}$ $\qquad \ldots\ldots(\gamma)$

Now take the logarithmic differential coefficient of U with respect to r, and we have

$$\frac{1}{U}\frac{dU}{dr} = -\frac{x^2+y^2-2rxy}{2}\frac{d}{dr}(1-r^2)^{-1}$$

$$-\frac{(1-r^2)^{-1}}{2}\frac{d}{dr}(x^2+y^2-2rxy) - \frac{1}{2}\frac{d}{dr}\log(1-r^2)$$

$$= -(x^2+y^2-2rxy)\, r(1-r^2)^{-2} + (1-r^2)^{-1}\, xy + r(1-r^2)^{-1}$$

$$\therefore \qquad (1-r^2)^2\frac{dU}{dr} = U\{xy + r(1-x^2-y^2) + r^2 xy - r^3\}$$

Differentiate n times by Leibnitz's theorem and put $r = 0$ and we have

$$u_{n+1} = n(2n - 1 - x^2 - y^2)\,u_{n-1} - n(n-1)\,(n-2)^2\,u_{n-3}$$
$$+ xy\{u_n + n(n-1)\,u_{n-2}\}$$

Hence
$$
\left.
\begin{aligned}
u_0 &= 1 \\
u_1 &= xy \\
u_2 &= (x^2 - 1)\,(y^2 - 1) \\
u_3 &= x(x^2 - 3)\,(y^2 - 3)\,y \\
u_4 &= (x^4 - 6x^2 + 3)\,(y^4 - 6y^2 + 3)
\end{aligned}
\right\} \qquad \ldots\ldots(\delta)
$$

The laws indicated by (δ) are

$$
\left.
\begin{aligned}
u_n &= v_n \times w_n \\
v_n &= xv_{n-1} - (n-1)\,v_{n-2} \\
w_n &= yw_{n-1} - (n-1)\,w_{n-2}
\end{aligned}
\right\} \qquad \ldots\ldots(\epsilon)
$$

and we can, therefore, re-write (β) as

$$\frac{1}{2\pi} U = \frac{1}{2\pi} e^{-\frac{1}{2}(x^2 + y^2)}\left(1 + \frac{v_1 w_1}{1!} r + \frac{v_2 w_2}{2!} r^2 + \ldots\right) \qquad \ldots\ldots(\zeta)$$

Integrating this from h to ∞ with respect to x, and remembering that w_n does not involve x, we have

$$\frac{1}{2\pi} \int_h^\infty U\,dx = \frac{1}{2\pi} e^{-\frac{1}{2}y^2}\left\{\int_h^\infty e^{-\frac{1}{2}x^2}dx + \frac{rw_1}{1!}\int_h^\infty e^{-\frac{1}{2}x^2} v_1\,dx + \ldots\right.$$

$$\left. + \frac{r^n w_n}{n!}\int_h^\infty e^{-\frac{1}{2}x^2} v_n\,dx + \ldots\right\}$$

$$= \frac{1}{\sqrt{(2\pi)}} e^{-\frac{1}{2}y^2}\left\{V_0 + \frac{rV_1 w_1}{1!} + \ldots + \frac{r^n V_n w_n}{n!} + \ldots\right\}$$

where V_n is written for $\dfrac{1}{\sqrt{(2\pi)}}\displaystyle\int_h^\infty v_n e^{-\frac{1}{2}x^2}dx$.

Now integrate this with respect to y from k to ∞, remembering that V_n does not involve y, and writing W_n for $\dfrac{1}{\sqrt{(2\pi)}}\displaystyle\int_h^\infty e^{-\frac{1}{2}y^2} w_n\,dy$,

we see that $\dfrac{1}{2\pi}\displaystyle\int_h^\infty \int_k^\infty U\,dx\,dy$ can be expressed as a series of

(247)

which the general term is $\dfrac{r^n}{n!}V_nW_n$, and we must now evaluate V_n and W_n.

From (ϵ) it can be shown by induction that the general form of v_n is

$$x^n - n_2 x^{n-2} + 3n_4 x^{n-4} - 5n_6 x^{n-6} + \ldots \qquad \ldots\ldots(\eta)$$

where $\qquad n_t = n(n-1)(n-2)\ldots(n-t+1)/t!$

Now we notice that

$$\frac{dv_n}{dx} = nv_{n-1} \qquad\qquad \ldots\ldots(\theta)$$

$\therefore$ by (ϵ) $\qquad\qquad v_n = xv_{n-1} - \dfrac{dv_{n-1}}{dx}$

Multiply by $e^{-\frac{1}{2}x^2}$ and integrate

$$\int e^{-\frac{1}{2}x^2} v_n dx = \int xe^{-\frac{1}{2}x^2} v_{n-1} dx - \int e^{-\frac{1}{2}x^2} \frac{dv_{n-1}}{dx} dx$$

and integrating the latter integral by parts, we have

$$\int e^{-\frac{1}{2}x^2} v_n dx = -e^{-\frac{1}{2}x^2} v_{n-1}$$

or $\qquad V_n = \dfrac{1}{\sqrt{(2\pi)}} \displaystyle\int_h^\infty v_n e^{-\frac{1}{2}x^2} dx = \dfrac{1}{\sqrt{(2\pi)}} e^{-\frac{1}{2}h^2}(v_{n-1})_{x=h}$

Now, writing H for $\dfrac{1}{\sqrt{(2\pi)}} e^{-\frac{1}{2}h^2}$, and K for $\dfrac{1}{\sqrt{(2\pi)}} e^{-\frac{1}{2}k^2}$, we have from ($\alpha$)

$$\frac{d}{N} = \frac{1}{2\pi} \int_h^\infty \int_k^\infty U dx\, dy$$

$$= \frac{1}{2\pi} \int_h^\infty \int_k^\infty e^{-\frac{1}{2}(x^2+y^2)} dx\, dy + \underset{1}{\overset{\infty}{S}}\left(\frac{r^n}{n!} HK(v_{n-1})_{x=h}(w_{n-1})_{y=k}\right)$$

$$= \frac{(b+d)(c+d)}{N^2} + \underset{1}{\overset{\infty}{S}}\left(\frac{r^n}{n!} HK(v_{n-1})_{x=h}(w_{n-1})_{y=k}\right)$$

or remembering that $N = a+b+c+d$, we write

$$\frac{ad-bc}{N^2HK} = \overset{\infty}{\underset{1}{S}} \left(\frac{r^n}{n!} (v_{n-1})_{x=h} \times (w_{n-1})_{y=k} \right)$$

$$= r + \frac{r^2}{2} hk + \frac{r^3}{6} (h^2 - 1)(k^2 - 1) + \frac{r^4}{24} h(h^2 - 3) k(k^2 - 3)$$

$$+ \frac{r^5}{120} (h^4 - 6h^2 + 3)(k^4 - 6k^2 + 3)$$

$$+ \frac{r^6}{720} h(h^4 - 10h^2 + 15) k(k^4 - 10k^2 + 15)$$

$$+ \frac{r^7}{5040} (h^6 - 15h^4 + 45h^2 - 15)(k^6 - 15k^4 + 45k^2 - 15)$$

$$+ \text{ etc. ...} \qquad\qquad\qquad ......\text{(viii)}$$

APPENDIX V

OTHER METHODS OF FITTING CURVES

The method of moments described in Chapter III is not the only method available for fitting curves. As would be expected of any general method, the method of moments may give difficulty because (i) the equations resulting from

$$\int_h^k x^n f(x, a, b, \ldots) \, dx = \nu_n$$

cannot be solved to find the constants $a, b, \ldots$ in $y = f(x, a, b, \ldots)$ or (ii) sufficiently accurate adjustments are not yet available in all circumstances for the statistical moments or (iii) some of the moments are liable to such large standard errors as to make them unreliable. This last point can be exemplified by curves, within the Pearson system, which cannot be fitted satisfactorily by moments, e.g. $y = y_0(1 + x^2)^{-1}$.

I. *The method of least squares.*

This method of adjusting observations makes the sum of the squares of the differences between the actual facts and the adjusted figures a minimum. The theory underlying the method is that errors are distributed in accordance with the "normal curve of error".

An example of the application of the method to fitting

$$a + bx + cx^2 + \ldots$$

where the observations are of equal weight would make

$$S\{y - (a + bx + cx^2 + \ldots)\}^2$$

a minimum where y stands for the observation. The equations for finding a, b, c, etc. would be

$$S(y) = aS(1) + bS(x) + cS(x^2) + \ldots$$
$$S(xy) = aS(x) + bS(x^2) + cS(x^3) + \ldots$$
$$S(x^2y) = aS(x^2) + bS(x^3) + cS(x^4) + \ldots$$

and so on until we have sufficient equations: then these equations have to be solved for a, b, c, etc.

The method has been extended to cases where a definite condition holds (e.g. the three angles of a triangle are measured and have to be adjusted subject to the condition that the total is equal to two right angles) or where observations have unequal weight. The reader who wishes to pursue this line of work may be referred to one of the text-books on the subject.*

The weakness of the method for fitting curves is that it often leads to equations which are incapable of solution. This can be seen by writing down and trying to solve equations similar to those set out above when $f(x)$ is taken as, say, a frequency curve of Type III instead of $a + bx + cx^2 + \ldots$. It follows that for general curve fitting the method of moments is preferable.† When the two methods can be compared conveniently they both give good results—probably, equally good. It is sometimes asserted that the method of least squares must give the best result. If the test of "best" is the smallest sum of the squares of the differences, then, obviously, the result is the best, but this is arguing in a circle. The difficulty lies in defining "best". A fair summary as regards curve fitting is that both methods give very good results but the method of moments is more generally applicable.

II. *The method of maximum likelihood.*

This is due to R. A. Fisher.‡ If a distribution is to be represented by $y = f(x, a, b, \ldots)$, where $a, b, \ldots$ are the unknown constants and the chance of an observation falling in the range dx is represented by $f(x, a, b, \ldots)\, dx$, then the chance that in a sample of n there will be n_1 in the range dx_1, n_2 in the range dx_2, and so on, is

$$\frac{n!}{\Pi(n_p!)}\, \Pi\{f(x_p, a, b, \ldots)\, dx_p\}^{n_p}$$

the products being taken for all values of p.

---

* E.g. D. Brunt's *Combination of Observations*. Cambridge, 1917.

† Thiele's method of half invariants (*Theory of Observations*) is allied to the method of moments rather than to the method of least squares.

‡ See *Philos. Trans.* A, ccxxii, 309 *et seq.*; *Proc. Camb. Phil. Soc.* xxii, 700 *et seq.*

The "method of maximum likelihood" consists in choosing values of a, b, ... which make this chance a maximum. As a, b, ... are only involved in f, we have to make

$$\Pi\{f(x_p, a, b, \ldots)\, dx_p\}^{n_p}$$

or

$$S\{n_p \log f(x_p, a, b, \ldots)\}$$

a maximum.

The only attempt I have seen to use the method in fitting curves such as those mainly dealt with in this book is with a Type I curve, and it would not be fair to judge the method by a single example. In that case the method began by using moments and then sought to improve the constants. The difficulty of using the method in general curve fitting is the same as that mentioned in connection with least squares; the equations reached cannot be solved directly and the constants have to be found by approximation.

Fisher's papers deal largely with other uses of "likelihood". For instance, he uses it to test whether the method of moments will give "efficient"* values for the constants and throws doubt on the method when the curve is far removed from the "Normal curve of error" type. The examples given in Chapters V and VI above include curves that may be considered open to this technical criticism. It would not appear, however, that the graduations are bad. Moreover, many other similar curves have been fitted by moments and good practical results have been obtained. As tentative suggestions, perhaps the explanation may be that we are dealing with constants that do not tend to normality even if we obtain large samples showing the distribution of those constants, or that larger samples than those ordinarily obtained for curve fitting are needed before the tendency to normality is reached, or that the constants are

* The criterion of "efficiency" is "that in large samples, when the distributions of the statistics tend to normality, that statistic is to be chosen that has the least probable error". Fisher uses "statistic" for a constant, or measure obtained from statistical samples: its plural is intended in the tenth word of the quotation.

not independent but correlated* or that the test of maximum likelihood is too refined to be of much help in practical curve fitting. At any rate, until some other better curve fitting process is available, there can be no harm in using a method that, in practice, fits curves satisfactorily to statistical facts.

III. *Minimum* χ^2.

This method was suggested by Dr Kirstine Smith† and fits curves by making χ^2 a minimum. This comes to the same thing as making $S\left(\dfrac{n_s^2}{m_s}\right)$ a minimum, where n is the statistical frequency and m the graduated.

Difficulty arises as in the methods described above because the resulting equations cannot be solved directly, and K. Smith suggested that the constants should first be calculated by moments and then adjusted by minimum χ^2. The additional work involved in any such double system of curve fitting would be justified if the improvement in fitting is considerable or if the problem demands refined treatment. The paper quoted gave some examples.

There has been a recent application of a minimum χ^2 method that was eminently practical. In a study of *Mortality Variations in Sweden*, H. Cramèr and H. Wold‡ assumed Makeham's law of mortality,§ $\mu_x = \alpha + \beta c^x$, and used

$$\chi_1^2 = \Sigma \frac{(\theta_x - \alpha E_x - \beta c^x E_x)^2}{\theta_x}$$

instead of

$$\chi^2 = \Sigma \frac{(\theta_x - E_x \mu_x)^2}{E_x \mu_x}$$

where E_x stands for the "Exposed to risk" of death at age x and θ_x for the corresponding deaths. Even with the modi-

* Anyone who does a large amount of curve fitting will have noticed that a mistake in working out one constant is not always wholly reflected in the curve. A simple example of what is meant can be inferred from the adjustment for J-shaped curves given in Appendix I.

† *Biometrika*, XI, 262 *et seq.*

‡ *Skandinavisk Aktuarietidskrift*, 1935, pp. 161 *et seq.*

§ Non-actuarial readers must not confuse the force of mortality, represented by μ_x in the international actuarial notation, with a moment.

fication, the equations are awkward. The first step was therefore to assume a definite value for c, and it was noticed that the methods of least squares and moments gave values for α and β which were equally good when tested by χ^2 and that a χ_1^2 test was also satisfactory. Trials were then made with a few values of c and the value of c finally chosen was such as to make χ^2 a minimum on the basis of passing a parabola through the values of χ^2. The following table gives an example of this part of the work:

1911–15. *Males. Values of χ^2, by method of moments. Graduation by parabolas of second and third degree.*

$10^3 \log c$	χ^2	Second degree parabola	Third degree parabola
43·25	40·02	39·968	40·020
43·50	32·09	32·143	32·091
43·75	27·41	27·461	27·409
44·00	25·92	25·921	25·921
44·25	27·58	27·525	27·577
44·50	32·32	32·271	32·323
44·75	40·11	40·111	40·109

The value of 44·00 was used for $10^3 \log c$ in this case. Incidentally it will be noticed from the table how great a change in χ^2 is caused by a comparatively small change in the value of $\log c$.

Having then decided on c, new values of β and α were calculated in that order and the α values by moments and by minimum χ^2 are shown.

As described, the method seems lengthy but it probably lent itself to systemization because graduations were in fact made from census particulars for quinquennia from 1801–5 to 1926–30 and for quinquennial ages from 32·5 to 87·5 for both sexes. The results were cross graduated so as to produce a smooth mortality surface for forecasting mortality. Similar tables were also produced for generations.

As an example of graduation on a large scale the paper should be studied both for its practical importance and interesting methods.

APPENDIX VI

KEY TO THE ACTUARIAL TERMS AND SYMBOLS USED

The following explanation of certain technical terms and symbols that are used in this book is given as an assistance to non-actuarial readers. For a fuller account of the functions and notation, reference can be made to the Text-Books for Actuaries *Interest and Annuities Certain*, by R. Todhunter and *Life Contingencies*, by E. F. Spurgeon. They are published by the Cambridge University Press.

When an investigation is made into the mortality experienced among lives assured, the number of persons entering at each age, and the numbers passing out of observation at each age owing to (1) *death*, (2) *withdrawal*, by the policies lapsing, being surrendered or terminating from some other cause, are recorded. The *exposed to risk* of death at age 25 (E_{25}) means the number who had the chance of dying between ages 25 and 26, and were on the average at risk for the whole year.

The number who die between ages 25 and 26, divided by the exposed to risk at age 25, gives *the probability of dying in a year* at age 25 (q_{25}). Consequently, deaths are tabulated according to age last birthday when the exposed to risk are tabulated according to the exact age or an approximation thereto; if, as is often necessary, the investigation is made according to the number of years an insurance has been in force, the exposed to risk are tabulated according to an exact duration and deaths according to the integral number of years in force (*curtate duration*).

When an experience ends in any year, say 1930, there will be a large number of persons who have been at risk, but whose policies are still in force; these are called *existing at the close of the observation*. When a graduation has been made the *expected deaths* are found by multiplying the exposed to risk by the graduated values of q_x. The result is then compared with the actual deaths.

$O^{NM(5)}$ is the name given to the table of mortality obtained from the male lives assured by ordinary whole-life without profit policies between 1863 and 1893, excluding the first five years of assurance.

O^M is a table constructed from the similar with-profit assurances for all durations, and H^M (healthy males) is the name given to the older experience which ended in 1863.

q_x is (see above) the probability of dying in a year.

p_x is the probability of a person aged x living one year.

So if we imagine a stationary community, which a person can only leave by death, and consider l_x to be the number living at exact age x, then $l_{x+1} = p_x \times l_x$ and $l_{x+2} = p_{x+1} \times l_{x+1}$, and so on.

The value, at a rate of interest i per unit, of a sum of 1 payable if a person aged x be alive at the end of n years, is therefore $\dfrac{v^n l_{x+n}}{l_x}$ where $v = (1+i)^{-1}$, and the value of an annuity of 1 would be

$$\frac{v l_{x+1} + v^2 l_{x+2} + \cdots}{l_x}$$

Now for convenience in making tables it is well to multiply numerator and denominator of this expression by v^x, and we have as the value of the annuity

$$\frac{v^{x+1} l_{x+1} + v^{x+2} l_{x+2} + \cdots}{v^x l_x} = \frac{D_{x+1} + D_{x+2} + \cdots}{D_x} = \frac{N_{x+1}}{D_x}$$

where $\qquad D_x = v^x l_x$ and $N_x = D_x + D_{x+1} + \cdots$

Similarly $\qquad S_x = N_x + N_{x+1} + \cdots$

Tables of D, N, and S are called *commutation* columns.

$a_{\overline{n}|}$ is written for the value of an annuity of 1 payable for n years certain, independent of any life, so its value is

$$v + v^2 + \cdots + v^n$$

$\bar{a}_{\overline{n}|}$ is the value of a similar annuity of 1 per annum payable m times a year when m takes the limit of ∞, so that its value is

$$\int_0^n v^m \, dm$$

Colog p_x is the logarithm of the reciprocal of p_x, and Makeham's hypothesis assumes that its value is $A + Bc^x$; but colog $p_x = \log l_x - \log l_{x+1}$; therefore an alternative way of stating the hypothesis is $l_x = k s^x g^{c^x}$.

The *force of mortality* (μ_x) is $-\dfrac{1}{l_x}\dfrac{dl_x}{dx}$. On Makeham's hypothesis it takes the form $\alpha + \beta c^x$.

When valuing the policies in an assurance office the actuary groups cases together to save labour. When assurances are payable at death they are grouped according to the year of birth, but when they are payable at a certain *maturity age* or previous death (*Endowment Assurances*) they can be grouped either according to year of birth or according to the number of years to run (*unexpired term*). In the latter case they are usually valued by finding an average age at maturity; formerly this was done by taking the mean of the ages, but G. J. Lidstone has shown that a more accurate result is reached by weighting the ages in Geometrical Progression. The constants used for this purpose are called Z.

A *model office* is an imaginary specimen office which is used for making approximate valuations.

APPENDIX VII

ABRIDGED READING

A student who needs to know little of the details of curve fitting and graduation but wants to have a general idea of what moments and frequency curves mean, should read Chapters I and II and the first eight articles of Chapter III, introducing the subject of fitting curves by moments (§§ 9–13 give an alternative arithmetical method and may be omitted). The student should understand that the arithmetical calculations of moments give a result that approximates to the corresponding moments found from the curve adopted for graduation (see §§ 15 and 16), but §§ 17–22 may be omitted. §§ 23–5 may be read, but little time need be spent on the algebra, and the adjustments of § 26 may be accepted as sufficient. § 27 is a general summary and should be read carefully.

Chapter IV may be abridged by reading § 1 and examining the table facing p. 51 (especially the "Remarks" column) and the folding diagram at the end of the book. These curves form Pearson's complete system: they express the frequency (y) and are evolved from the differential equation

$$\frac{d \log y}{dx} = \frac{x + a}{b_0 + b_1 x + b_2 x^2}$$

Chapter V shows how each type can be fitted by the method of moments. All the mathematical part may be omitted, but the numerical examples distributed through the chapter should be examined in order to form a proper idea of the range of applicability of the system of curves. Pp. 80–85 should be read; they deal with the "normal curve of error" and supply a simple example of curve fitting. Chapter VI should be read, omitting the processes of § 5; it gives a comparison of various systems of curves.

Chapter VII opens the subject of correlation: it is unnecessary at an early stage to use more than one arithmetical method of calculating the coefficient of correlation and §§ 11–13 may be omitted. The remaining chapters should be read, but a non-mathematical reader who is content to appreciate the common-sense of results without following all the algebraic treatment, may omit the algebra on pp. 161–4. The Appendices except V may be omitted.

APPENDIX VIII

REFERENCES, ETC.

For working out examples similar to those in this book, it is necessary to use a 7-figure table of logarithms, trigonometric functions, etc., such as *Chambers'* and Barlow's *Tables of Squares, etc.* A multiplying machine will save much heavy arithmetic.

The following will be found of help in various ways:

Tables for Statisticians and Biometricians, Parts I and II. Edited by K. Pearson. Camb. Univ. Press.

Tracts for Computers. Camb. Univ. Press, especially IX, giving $\log \Gamma(x)$ from $x = 1$ to $50\cdot9$ by intervals of $\cdot01$; XII, *Tables of the Probable Error of the Coefficient of Correlation*; and XV, *Random Sampling Numbers.*

The following lists of books and papers deal with the subject generally or with parts of it. The lists are confined to work written in English and do not aim at a complete bibliography. The books in the first list deal with parts of the subject from various points of view and will be found useful for further reading. The lists of papers include those mentioned in the text but do not give all the authorities consulted. Papers are not always included when their contents are more conveniently read in one of the books (e.g. in Fisher's *Statistical Methods* or in the *Introductions to Tables for Statisticians*). The division between subjects is necessarily somewhat arbitrary.

Most of the original work in English on statistical mathematics has appeared in the *Philosophical Transactions* or the *Proceedings of the Royal Society,* the *Transactions of the Royal Society of Edinburgh,* or in such journals as *Biometrika, Metron, Journal of Royal Statistical Society, Annals of Eugenics, Annals of Mathematical Statistics* and on actuarial subjects in the *Journals* of the Institute of Actuaries and the

Faculty of Actuaries or in *Skandinavisk Aktuarietidskrift*. A student who wishes to trace the past development of the subject or keep in touch with future advances may be referred to them.

LIST OF BOOKS

A. L. Bowley. *Elements of Statistics*. P. S. King & Son. *F. Y. Edgeworth's Contributions to Mathematical Statistics*. Royal Statistical Society.

W. Brown and G. H. Thomson. *The Essentials of Mental Measurement*. Camb. Univ. Press.

D. Brunt. *The Combination of Observations*. Camb. Univ. Press.

R. A. Fisher. *Statistical Methods for Research Workers*. Oliver & Boyd. *The Design of Experiments*. Oliver & Boyd.

D. C. Jones. *First Course in Statistics*. G. Bell & Sons.

T. L. Kelley. *Statistical Method*. The Macmillan Co.

E. S. Pearson. *Application of Statistical Methods to Industrial Standardisation and Quality Control*. British Standards Institution.

H. E. Soper. *Frequency Arrays*. L. Reeve & Co., Ltd.

J. F. Steffensen. *Some Recent Researches in the Theory of Statistics and Actuarial Science*. Camb. Univ. Press.

T. N. Thiele. *Theory of Observations*. London 1903.

G. H. Thomson. *How to Calculate Correlations*. G. G. Harrap & Co.

L. H. C. Tippett. *Methods of Statistics*. Williams & Norgate.

E. T. Whittaker and G. Robinson. *Calculus of Observations*. Blackie and Co.

G. U. Yule and M. G. Kendall. *Introduction to the Theory of Statistics*. Charles Griffin & Co.

LISTS OF PAPERS

Method of Moments, etc.

H. Cramèr and H. Wold. "Mortality Variations in Sweden." *Skand. Aktuarietidskrift*, 1935, p. 161.

R. A. Fisher. "On the mathematical foundations of theoretical statistics." *Philos. Trans.* A, CCII, p. 309. See also *Proc. Camb. Phil. Soc.* XXII, p. 700.

E. S. Martin. "On corrections for the moment coefficients of frequency distributions when the start of the frequency is one of the characteristics to be determined." *Biometrika*, XXVI, p. 12.

E. Pairman and K. Pearson. "Corrections for moment coefficients." *Biometrika*, XII, p. 231.

K. Pearson. "Systematic fitting of curves to observations." *Biometrika*, I, p. 265; II, p. 1.

W. F. Sheppard. "Calculation of the most probable values of frequency constants, etc." *Proc. Lond. Math. Soc.* XXIX, p. 353.

K. Smith. "On the 'Best' values of the constants in Frequency Distributions." *Biometrika*, XI, p. 262.

(261)

Frequency-curves

C. V. L. Charlier. "Researches into the theory of probability." *Meddelanden Lunds Astronomiska Observatorium*, 1906. "A new form of the Frequency Function." *Ibid.* 1928.

H. Cramèr. "On the composition of elementary errors." *Skand. Aktuarietidskrift*, 1928, pp. 13 and 141.

F. Y. Edgeworth. "Mathematical representation of statistical data." *J. Roy. Statist. Soc.* LXI, p. 691; LXIX, p. 497; LXX, p. 102; LXXIX, p. 455; LXXX, pp. 65, 266, 411; LXXXVII, p. 571. "Generalised Law of Error." *Proc. Camb. Phil. Soc.* XX, pp. 36, 113.

W. P. Elderton. "An approximate law of survivorship and other notes on the use of frequency curves in actuarial statistics." *J. Inst. Actu.* LXV, p. 1.

J. C. Kapteyn. *Skew Frequency Curves in Biology and Statistics.* Groningen, 1903 and 1916.

K. Pearson. "Skew variation in homogeneous material." *Philos. Trans.* A, CLXXXVI, p. 343; CXCVII, p. 443; CCXVI, p. 429.

E. C. Rhodes. "On the generalised law of error." *J. Roy. Statist. Soc.* LXXXVIII, p. 576.

Correlation and Contingency

E. M. Elderton. "Relative value of factors influencing infant welfare." *Annals of Eugenics*, I.

F. Galton. "Family likeness in Stature." *Proc. Roy. Soc.* XI, p. 42. "Correlations and their measurement." *Proc. Roy. Soc.* XLV, p. 136.

K. Pearson. "Regression, Heredity and Panmixia." *Philos. Trans.* A, CLXXXVII, p. 253. "Correlation of characters not quantitatively measurable." *Philos. Trans.* A, CXCV, p. 1. "Theory of Contingency; Theory of Skew Correlation; Further methods of measuring correlation; Novel method of regarding the association of two variates." *Drapers' Company Memoirs*, Camb. Univ. Press. "New method of determining Correlation." *Biometrika*, VII, p. 248. "On a form of spurious correlation." *Proc. Roy. Soc.* LX, p. 489. Also *Biometrika*, VIII, p. 254; IX, p. 116; XIV, p. 281.

C. Spearman. "Proof and measurement of Association between two things." *Amer. J. Psychol.* XV, p. 72; also *Brit. J. Psychol.* II, p. 96.

"Student." "Elimination of Spurious Correlation, etc." *Biometrika*, X, p. 179.

G. U. Yule. "Significance of Bravais' Formulae, etc." *Proc. Roy. Soc.* LX, p. 477. "Theory of Correlation." *J. Roy. Statist. Soc.* LX, p. 812. "On time correlation problem." *J. Roy. Statist. Soc.* XXXIV, p. 497; also *J. Roy. Statist. Soc.* LXXXIX, p. 1.

Standard Errors, Goodness of Fit, etc.

Biometrika, editorial. "Probable errors of frequency constants." *Biometrika*, II, p. 273.

F. L. Engledow and G. U. Yule. *Principles and Practice of Yield Trials*. Empire Cotton Growing Corporation, 1926.

R. A. Fisher. Various papers dealing with the subject and especially with χ^2 and small samples. *J. Roy. Statist. Soc.* LXXXV, p. 597; LXXXVII, p. 442; *Economica*, 1923, p. 139; *Biometrika*, X, p. 507; *Metron*, I, pt. 4, p. 1; V, pt. 3, p. 92.

R. Henderson. "Frequency curves and moments." *Trans. Actu. Soc. America*, VIII, p. 30 or *J. Inst. Actu.* XLI, p. 429.

M. Greenwood. "Errors in random sampling." *Biometrika*, IX, p. 69.

T. Kondo. "Standard error of mean square contingency." *Biometrika*, XXI, p. 376.

E. S. Pearson. "Some notes on sampling tests with two variables." *Biometrika*, XXI, p. 337.

K. Pearson. "Criterion that a given system of deviations can have arisen from Random Sampling." *Phil. Mag.* July 1900. See also various other papers. *Biometrika*, VIII, p. 250; XI, p. 292; XIV, pp. 186, 418. (With L. N. G. Filon.) *Philos. Trans.* A, CXCI, p. 229.

"Student." "Small samples." *Biometrika*, VI, p. 302; X, p. 384; XV, p. 271. "Ranks and grades correlation method." *Biometrika*, XIII, p. 264.

J. Wishart and H. G. Sanders. *Principles and Practice of Field Experimentation*. Empire Cotton Growing Corporation.

G. U. Yule. "Application of χ^2 to contingency tables." *J. Roy. Statist. Soc.* LXXXV, p. 95.

APPENDIX IX

TABLES

Useful Constants

$e = 2{\cdot}71828\ 18285$

$e^{-1} = {\cdot}36787\ 94417$

$\pi = 3{\cdot}14159\ 26536$

$\log_{10} e = {\cdot}43429\ 44820$

$\log_e 10 = 2{\cdot}30258\ 509$

$\log(\log_{10} e) = \overline{1}{\cdot}63778\ 43114$

$\log_{10} \pi = {\cdot}49714\ 98728$

$\log_{10} \sqrt{\pi} = {\cdot}24857\ 49364$

$\log_{10} \dfrac{1}{\sqrt{(2\pi)}} = \overline{1}{\cdot}60091\ 00657$

$\log_{10} e^{-\frac{1}{2}} = \overline{1}{\cdot}96380\ 87932$

Areas and ordinates of Normal Curve of Error in terms of abscissa

x	$\frac{1}{2}(1 + \alpha)$	z
·0	·5000	·3989
·1	·5398	·3970
·2	·5793	·3910
·3	·6179	·3814
·4	·6554	·3683
·5	·6915	·3521
·6	·7257	·3332
·7	·7580	·3123
·8	·7881	·2897
·9	·8159	·2661
1·0	·8413	·2420
1·1	·8643	·2179
1·2	·8849	·1942
1·3	·9032	·1714
1·4	·9192	·1497
1·5	·9332	·1295
1·6	·9452	·1109
1·7	·9554	·0940
1·8	·9641	·0790
1·9	·9713	·0656
2·0	·9772	·0540
2·5	·9938	·0175
3·0	·9987	·0044
3·5	·9998	·0009

$z = \dfrac{1}{\sqrt{(2\pi)}\sigma}\, e^{-\frac{1}{2}(x/\sigma)^2}$. The tabulation assumes $\sigma = 1$.

$\alpha = \displaystyle\int_{-x}^{x} z\, dx$ or $\tfrac{1}{2}(1 + \alpha) = \displaystyle\int_{-\infty}^{x} z\, dx$.

Test for Goodness of Fit. Values of P.

χ^2	$n = 2$ $n' = 3$	$n = 5$ $n' = 6$	$n = 8$ $n' = 9$	$n = 11$ $n' = 12$	$n = 14$ $n' = 15$	$n = 17$ $n' = 18$	$n = 20$ $n' = 21$	$n = 23$ $n' = 24$	$n = 26$ $n' = 27$	$n = 29$ $n' = 30$	χ^2
3	·223	·700	·934	·991	·999	1·000	1·000	1·000	1·000	1·000	3
6	·050	·306	·647	·873	·966	·993	·999	1·000	1·000	1·000	6
9	·011	·109	·342	·622	·831	·940	·983	·996	·999	1·000	9
12	·002	·035	·151	·363	·606	·800	·916	·970	·991	·998	12
15	·001	·010	·059	·182	·378	·595	·776	·895	·957	·985	15
18	·000	·003	·021	·082	·207	·389	·589	·757	·876	·944	18
21	·000	·001	·007	·033	·102	·226	·397	·581	·742	·859	21
24	·000	·000	·002	·013	·046	·119	·243	·404	·576	·729	24
27	·000	·000	·001	·005	·019	·058	·135	·256	·409	·572	27
30	·000	·000	·000	·002	·008	·026	·070	·149	·268	·414	30

If $\chi^2 = 0$, $P = 1$. Rough interpolations can be made either horizontally or vertically, but not diagonally, by using log P instead of P. n is the number of "degrees of freedom" and $n' = n + 1$ the basis of tabulation in *Tables for Statisticians*.

Table of log $\Gamma(p)$

p	0	1	2	3	4	5	6	7	8	9
1·00	...	...	9750	9500	9251	9003	8755	8509	8263	8017	7773
1·01	1̄·99	7529	7285	7043	6801	6560	6320	6080	5841	5602	5365
1·02	1̄·99	5128	4892	4656	4421	4187	3953	3721	3489	3257	3026
1·03	1̄·99	2796	2567	2338	2110	1883	1656	1430	1205	0981	0757
1·04	1̄·99	0533	0311	0089	9̄868	9̄647	9̄427	9̄208	8̄989	8̄772	8̄554
1·05	1̄·98	8338	8122	7907	7692	7478	7265	7053	6841	6629	6419
1·06	1̄·98	6209	6000	5791	5583	5376	5169	4963	4758	4553	4349
1·07	1̄·98	4145	3943	3741	3539	3338	3138	2939	2740	2541	2344
1·08	1̄·98	2147	1951	1755	1560	1365	1172	0978	0786	0594	0403
1·09	1̄·98	0212	0022	9̄833	9̄644	9̄456	9̄269	9̄082	8̄896	8̄710	8̄525
1·10	1̄·97	8341	8157	7974	7791	7610	7428	7248	7068	6888	6709
1·11	1̄·97	6531	6354	6177	6000	5825	5650	5475	5301	5128	4955
1·12	1̄·97	4783	4612	4441	4271	4101	3932	3764	3596	3429	3262
1·13	1̄·97	3096	2931	2766	2602	2438	2275	2113	1951	1790	1629
1·14	1̄·97	1469	1309	1150	0992	0835	0677	0521	0365	0210	0055
1·15	1̄·96	9901	9747	9594	9442	9290	9139	8988	8838	8688	8539
1·16	1̄·96	8390	8243	8096	7949	7803	7658	7513	7369	7225	7082
1·17	1̄·96	6939	6797	6655	6514	6374	6234	6095	5957	5818	5681
1·18	1̄·96	5544	5408	5272	5137	5002	4868	4734	4601	4469	4337
1·19	1̄·96	4205	4075	3944	3815	3686	3557	3429	3302	3175	3048
1·20	1̄·96	2922	2797	2672	2548	2425	2302	2179	2057	1936	1815
1·21	1̄·96	1695	1575	1456	1337	1219	1101	0984	0867	0751	0636
1·22	1̄·96	0521	0407	0293	0180	0067	9̄955	9̄843	9̄732	9̄621	9̄511
1·23	1̄·95	9401	9292	9184	9076	8968	8861	8755	8649	8544	8439
1·24	1̄·95	8335	8231	8128	8025	7923	7821	7720	7620	7520	7420
1·25	1̄·95	7321	7223	7125	7027	6930	6834	6738	6642	6547	6453
1·26	1̄·95	6359	6267	6173	6081	5989	5898	5807	5716	5627	5537
1·27	1̄·95	5449	5360	5273	5185	5099	5013	4927	4842	4757	4673
1·28	1̄·95	4589	4506	4423	4341	4259	4178	4097	4017	3938	3858
1·29	1̄·95	3780	3702	3624	3547	3470	3394	3318	3243	3168	3094
1·30	1̄·95	3020	2947	2874	2802	2730	2659	2588	2518	2448	2379
1·31	1̄·95	2310	2242	2174	2106	2040	1973	1907	1842	1777	1712
1·32	1̄·95	1648	1585	1522	1459	1397	1336	1275	1214	1154	1094
1·33	1̄·95	1035	0977	0918	0861	0803	0747	0690	0634	0579	0524
1·34	1̄·95	0470	0416	0362	0309	0257	0205	0153	0102	0051	0001
1·35	1̄·94	9951	9902	9853	9805	9757	9710	9663	9617	9571	9525
1·36	1̄·94	9480	9435	9391	9348	9304	9262	9219	9178	9136	9095
1·37	1̄·94	9054	9015	8975	8936	8898	8859	8822	8785	8748	8711
1·38	1̄·94	8676	8640	8605	8571	8537	8503	8470	8437	8405	8373
1·39	1̄·94	8342	8311	8280	8250	8221	8192	8163	8135	8107	8080
1·40	1̄·94	8053	8026	8000	7975	7950	7925	7901	7877	7854	7831
1·41	1̄·94	7808	7786	7765	7744	7723	7703	7683	7664	7645	7626
1·42	1̄·94	7608	7590	7573	7556	7540	7524	7509	7494	7479	7465
1·43	1̄·94	7451	7438	7425	7413	7401	7389	7378	7368	7358	7348
1·44	1̄·94	7338	7329	7321	7312	7305	7298	7291	7284	7278	7273
1·45	1̄·94	7268	7263	7259	7255	7251	7248	7246	7244	7242	7241
1·46	1̄·94	7240	7239	7239	7240	7241	7242	7243	7245	7248	7251
1·47	1̄·94	7254	7258	7262	7266	7271	7277	7282	7289	7295	7302
1·48	1̄·94	7310	7317	7326	7334	7343	7353	7363	7373	7384	7395
1·49	1̄·94	7407	7419	7431	7444	7457	7471	7485	7499	7514	7529
p	0	1	2	3	4	5	6	7	8	9

Table of log $\Gamma(p)$—continued

p		0	1	2	3	4	5	6	7	8	9
1·50	Ī·94	7545	7561	7577	7594	7612	7629	7647	7666	7685	7704
1·51	Ī·94	7724	7744	7764	7785	7806	7828	7850	7873	7896	7919
1·52	Ī·94	7943	7967	7991	8016	8041	8067	8093	8120	8146	8174
1·53	Ī·94	8201	8229	8258	8287	8316	8346	8376	8406	8437	8468
1·54	Ī·94	8500	8532	8564	8597	8630	8664	8698	8732	8767	8802
1·55	Ī·94	8837	8873	8910	8946	8983	9021	9059	9097	9135	9174
1·56	Ī·94	9214	9254	9294	9334	9375	9417	9458	9500	9543	9586
1·57	Ī·94	9629	9672	9716	9761	9806	9851	9896	9942	9989	0035
1·58	Ī·95	0082	0130	0177	0225	0274	0323	0372	0422	0472	0522
1·59	Ī·95	0573	0624	0676	0728	0780	0833	0886	0939	0993	1047
1·60	Ī·95	1102	1157	1212	1268	1324	1380	1437	1494	1552	1610
1·61	Ī·95	1668	1727	1786	1845	1905	1965	2025	2086	2147	2209
1·62	Ī·95	2271	2333	2396	2459	2522	2586	2650	2715	2780	2845
1·63	Ī·95	2911	2977	3043	3110	3177	3244	3312	3380	3449	3517
1·64	Ī·95	3587	3656	3726	3797	3867	3938	4010	4081	4154	4226
1·65	Ī·95	4299	4372	4446	4519	4594	4668	4743	4819	4894	4970
1·66	Ī·95	5047	5124	5201	5278	5356	5434	5513	5592	5671	5740
1·67	Ī·95	5830	5911	5991	6072	6154	6235	6317	6400	6482	6566
1·68	Ī·95	6649	6733	6817	6901	6986	7072	7157	7243	7322	7416
1·69	Ī·95	7503	7590	7678	7766	7854	7943	8032	8122	8211	8301
1·70	Ī·95	8391	8482	8573	8664	8756	8848	8941	9034	9127	9220
1·71	Ī·95	9314	9409	9502	9598	9693	9788	9884	9980	0̄077	0̄174
1·72	Ī·96	0271	0369	0467	0565	0664	0763	0862	0961	1061	1162
1·73	Ī·96	1262	1363	1464	1566	1668	1770	1873	1976	2079	2183
1·74	Ī·96	2287	2391	2496	2601	2706	2812	2918	3024	3131	3238
1·75	Ī·96	3345	3453	3561	3669	3778	3887	3996	4105	4215	4326
1·76	Ī·96	4436	4547	4659	4770	4882	4994	5107	5220	5333	5447
1·77	Ī·96	5561	5675	5789	5904	6019	6135	6251	6367	6484	6600
1·78	Ī·96	6718	6835	6953	7071	7189	7308	7427	7547	7666	7787
1·79	Ī·96	7907	8028	8149	8270	8392	8514	8636	8759	8882	9005
1·80	Ī·96	9129	9253	9377	9501	9626	9751	9877	0̄003	0̄129	0̄255
1·81	Ī·97	0383	0509	0637	0765	0893	1021	1150	1279	1408	1538
1·82	Ī·97	1668	1798	1929	2060	2191	2322	2454	2586	2719	2852
1·83	Ī·97	2985	3118	3252	3386	3520	3655	3790	3925	4061	4197
1·84	Ī·97	4333	4470	4606	4744	4881	5019	5157	5295	5434	5573
1·85	Ī·97	5712	5852	5992	6132	6273	6414	6555	6697	6838	6980
1·86	Ī·97	7123	7266	7408	7552	7696	7840	7984	8128	8273	8419
1·87	Ī·97	8564	8710	8856	9002	9149	9296	9443	9591	9739	9887
1·88	Ī·98	0036	0184	0333	0483	0633	0783	0933	1084	1234	1386
1·89	Ī·98	1537	1689	1841	1994	2147	2299	2453	2607	2761	2915
1·90	Ī·98	3069	3224	3379	3535	3690	3846	4003	4159	4316	4474
1·91	Ī·98	4631	4789	4947	5105	5264	5423	5582	5742	5902	6062
1·92	Ī·98	6223	6383	6544	6706	6867	7029	7192	7354	7517	7680
1·93	Ī·98	7844	8007	8171	8336	8500	8665	8830	8996	9161	9327
1·94	Ī·98	9494	9660	9827	9995	0̄162	0̄330	0̄498	0̄666	0̄835	1̄004
1·95	Ī·99	1173	1343	1512	1683	1853	2024	2195	2366	2537	2709
1·96	Ī·99	2881	3054	3227	3399	3573	3746	3920	4094	4269	4443
1·97	Ī·99	4618	4794	4969	5145	5321	5498	5674	5851	6029	6206
1·98	Ī·99	6384	6562	6740	6919	7098	7277	7457	7637	7817	7997
1·99	Ī·99	8178	8359	8540	8722	8903	9085	9268	9450	9633	9816
p		0	1	2	3	4	5	6	7	8	9

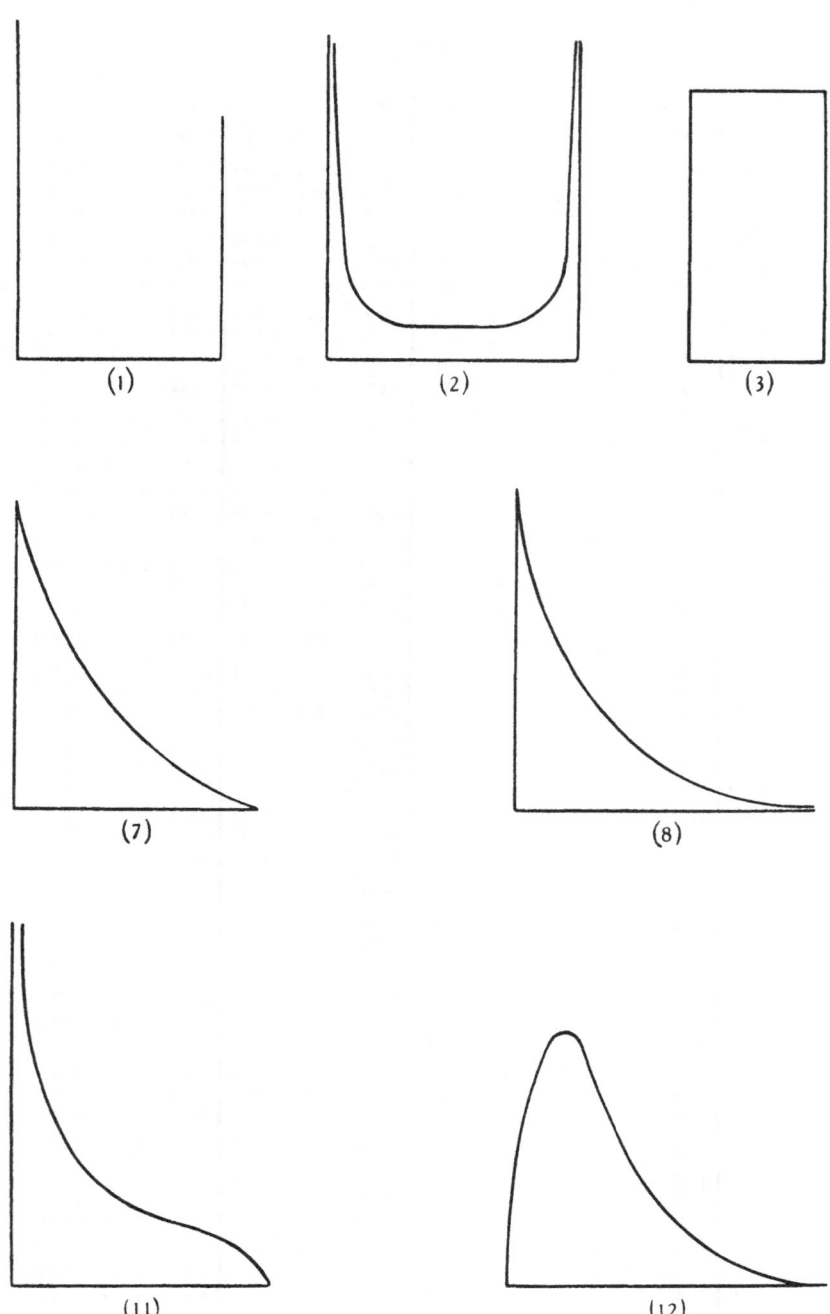

Diagram showing transition from two separated blocks of frequency through
U-shaped and J-shaped curves to the more common "cocked hat" shapes.

(1) represents two separated blocks of frequency developing into the U-shaped
curve in (2). The horizontal straight line of (3) is, as it were, the bottom piece of
the U-curve and the Type VIII curve of (4) is like a part of the U-curve. (5) and (6)

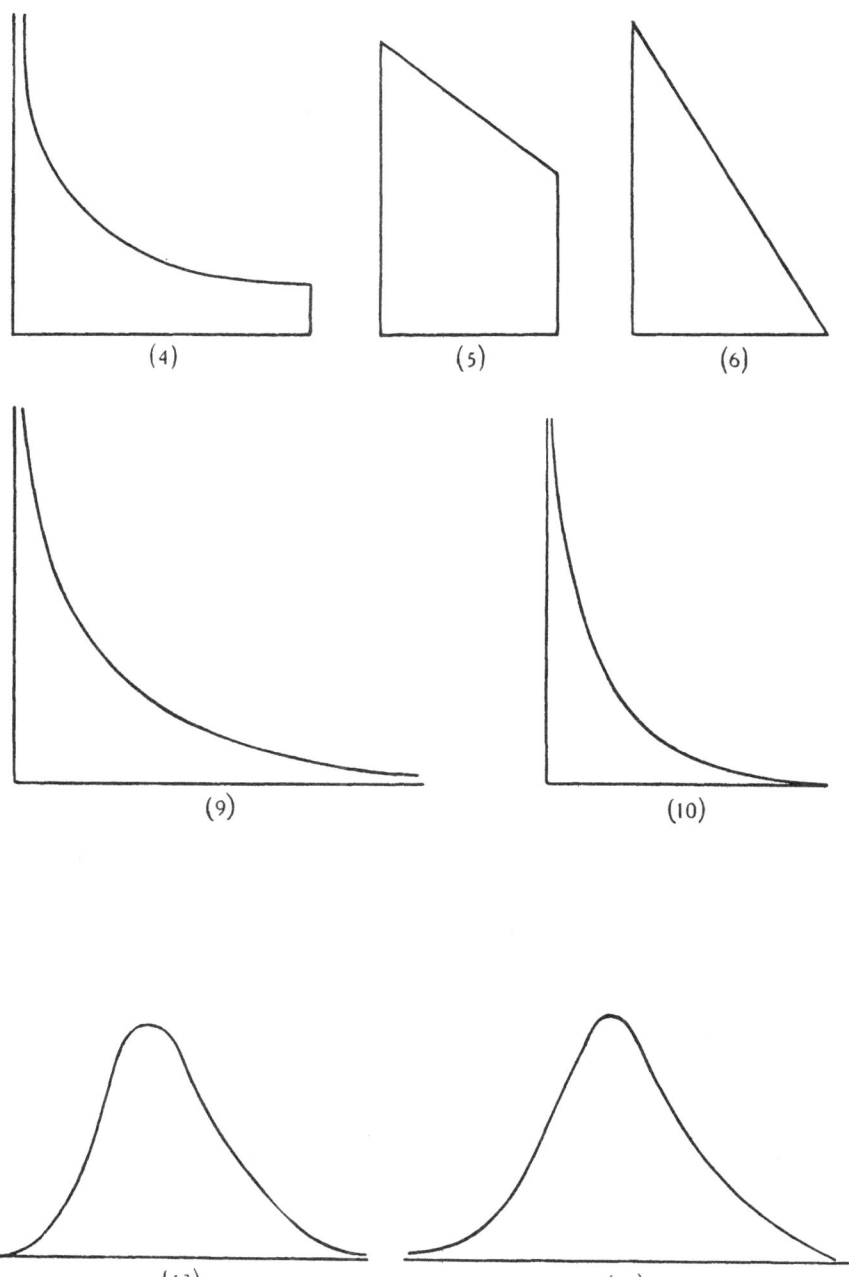

are limits when straight lines are reached. (7) is Type IX and (8) is the exponential. The next two curves (9) and (10) are the J-shaped curves of Types III and I, and (11) is Type XII. From this we proceed to Types I, III, IV, V and VI, curves of the "cocked hat" shape, three examples being given in (12), (13) and (14).

INDEX OF SUBJECTS

(*See also References in Appendix VIII*)